THE SABBATICAL

A Year of Travel During the Pandemic

James R. A. Herriot

An epic travel Odyssey compiled in lockdown which encompasses thought-provoking photography of 52 diverse locations; in addition, 52 books having made an impression, plus 52 much-loved classical and contemporary pieces of music, also 52 fine wines consumed with family and friends at the various destinations. A Grandfather's attempt to broaden his Children and Grandchildren's horizons and impart a message.

THE
SABBATICAL

A Year of Travel During the Pandemic

James R. A. Herriot

The Sabbatical: A Year of Travel During the Pandemic by James R. A. Herriot.

First edition published in Great Britain in 2023 by Extremis Publishing Ltd., Suite 218, Castle House, 1 Baker Street, Stirling, FK8 1AL, United Kingdom.

www.extremispublishing.com

Extremis Publishing is a Private Limited Company registered in Scotland (SC509983) whose Registered Office is Suite 218, Castle House, 1 Baker Street, Stirling, FK8 1AL, United Kingdom.

A CIP catalogue record for this book is available from the British Library.

ISBN: 978-1-7398543-6-2

Typeset in Linux Libertine.

Printed and bound in Great Britain by IngramSpark, Chapter House, Pitfield, Kiln Farm, Milton Keynes, MK11 3LW, United Kingdom.

THE
SABBATICAL

A Year of Travel During the Pandemic

James R. A. Herriot

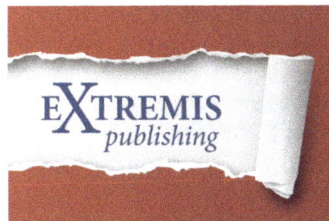

EXTREMIS
publishing

ACKNOWLEGEMENTS

I would like to acknowledge the input and efforts of 5 people, without whom the 'Sabbatical' would never have been completed and published. So a massive thank you to each and every one of you:

The technical and graphic input, as well as the encouragement and support, of my friend and work colleague David Knaggs has been pivotal – without this, I don't believe the 'Sabbatical' would have come to fruition. I cannot thank you enough for all your efforts; it's much appreciated.

Granny played an important role in proof-reading each individual weekly entry as I compiled my on-line journal during the 2nd COVID-19 lockdown. She kept me right as to names, dates, locations, etc., which ensured my facts and recollections were correct. Your help and assistance was invaluable.

Our No.1 Son, Simon (the fine wine specialist), played a critical role in helping me compile the list of wines. His passion and ability to remember wines we've consumed together, or ones I'd emailed details from faraway places, was critical in enabling this important element of the 'Sabbatical'.

To Julie and Tom Christie, the founders of Extremis Publishing: Without your belief and support in the 'Sabbatical' project, as well as your knowledge and understanding of publishing, my on-line journal would not now be in print. Thank you both; your help and assistance has played an essential role.

THE IDEA

The Sabbatical chronicles
a time of uncertainty and
an attempt via imagery, places,
words, people, memories,
composers, books, musicians,
and fine wine
to provide an antidote.

Our message:

The World is Your Oyster: do with it what you can... with an open and inquiring mind.

Dedicated to:

My long suffering
wife of 45 years,
our two much-loved children,
and four adored grandchildren.

*When reading the Sabbatical,
I suggest you listen to the music
assigned to each individual entry
to gain an insight into my mood
and thought when compiling
these.*

LETTER TO THE GRANDCHILDREN

To Our Grandchildren

Eight years, add in COVID-19 and a 2nd national lockdown, has finally driven me to collate my thoughts, put pen to paper (finger to keyboard), and bring the 'Sabbatical' to life. This may come across as a bit of a riddle - but there is a modicum of logic involved. The time-line of eight years is easy to remember, as this relates to our first Grandchild (JJ) coming into our lives. An amazing, joyous and momentous occasion - one never to be forgotten.

Around the same time as JJ was born, our family lost my brother-in-law David to a brain tumour, my cousin Tim to pancreatic cancer, followed within a short period of time by my cousin Pete (Tim's younger brother), also to pancreatic cancer. Three amazing people who achieved much in life.

The 'Sabbatical' came about due to Tim's funeral, which was attended by Pete and his loving family. Tim, Pete and I went back a long way, and a story in its own right – they were very dear to me. After the church service came the committal and I watched Pete, his wife and three children bury his older brother – knowing the funeral they could well attend next, may be his.

The funeral concluded at a pub in (North Yorkshire), during which time Pete and I had a long chinwag about Tim, our parents, and many great times gone by. The tear-jerker at the end being Pete's comment, "I'll see you soon," followed by a pause and "Maybe not". Pete, ever the pragmatist, knew time was not on his side. I was never to see Pete again.

Those fateful few months led me to dwell on how I'd never known any of my grandparents – all died before I was born, and I've little knowledge, background or understanding of them. What if I was to die? What would my new-born Grandson (JJ) know about me, his Grandpa? Hence the concept of the 'Sabbatical' came about.

A simple explanation of the 'Sabbatical' relates to 52 places in the world that Granny and I have visited and loved, each highlighted and demonstrated via pictures I've taken at the various locations.

To accompany the 52 places: 52 composers and pieces of classical music I treasure, followed by 52 musicians and songs recalling occasions and memories, 52 authors and books I've enjoyed that broadened my horizons, and the final element: 52 fine wines consumed in the company of family and friends in the 52 places.

I'm sure the question forming in your brain must be: "How will that help young JJ understand his Grandpa"?

The simple answer: If JJ were to visit the 52 places, listen to 52 composers, enjoy 52 great songs, learn from 52 books, and drink 52 fine wines, he may well know diddly-squat about his Granny and Grandpa – but he'd have one hell of time, and what a journey. And, just maybe, he would understand and gain an insight into his grandparents - who knew (still do) a thing or two about enjoying themselves. A valuable lesson in life.

So for JJ, as well as Lozzie and baby Sandy (we've since been blessed with 2 additional grandchildren) – when the time is right, you may well contemplate embarking on the 'Sabbatical'. Current circumstances from Granny and Grandpa's point of view, due to the imposition of a global pandemic and lockdown, means travel is out for now.

But that doesn't stop us from sharing our memories of places, music, books and fine wines with you and others.

Enjoy – Grandpa

WEEK 1
THE LAXFORD SCOTLAND

2020 should have been the 30th anniversary since I first visited one of my favourite places on the planet. The global pandemic well and truly scuppered that. The Scottish Highlands are special and incredibly dear to me - in particular, my annual trip to fish the Laxford. The place is rugged, beautiful, ever so peaceful, and just good for the soul.

It's been an honour and privilege to have had the opportunity to visit such an amazing place over many years.

My love affair with the area came about due to an invitation from my Dad - to join him on a salmon fishing trip to Sutherland. I'd not fished since a child, and decided to go, simply for him. One week and 8 salmon later, I was completely hooked. Not only with salmon fishing, the beauty of the place, but also the amazing people I've been lucky enough to get to know and call my friends.

That first visit transformed my outlook on Scotland and the Highlands.

In the early days we stayed at Stack Lodge, which looks up the length of Loch Stack. To the left is Arkle, to the right Ben Stack, and in the distance Foinaven. I've seen the loch in many spectacular guises – from flat calm, as per my picture with Arkle reflected in the water, to one amazing night where the stars were mirrored in the loch like a planetarium. Others where the waves have been enormous, and water spouts hurtling across the loch.

It's difficult to explain how beautiful and brutal the weather can be – all in one day. I've come back from the river for lunch where the sun is beating down; and encountered torrents of water and landslides coming off the side of the hills on my return to the river. Experiencing four seasons in a day is not that unusual.

One memorable day while fishing with the legendary Willie (The Gillie), Dad, having more sense, called off, as the rain was torrential and passing the windows horizontally - the wind best described as wild and violent. Having fished all morning and just extracted a large tube-fly from my forehead, and nothing to show for it other than blood, we decided one last try before lunch. Bingo! Into a fish and landed, while doing a jig in the storm to celebrate.

Willie shouts in my ear: "If you see any men in white coats – make a run for it"!

13

In more recent times we've stayed at the Old Laundry in Achfary, with the burn running alongside the house. Great excitement when waking in the morning and hearing the roar of water – bodes well for the day's fishing. The view from the house up Loch More, on a clear day or early evening, is special and spectacular. The food and hospitality from Trish and team is outstanding.

A very special place to visit with one's friends.

The river Laxford runs for 4 miles from Loch Stack to the sea. For me, it's magical; all its many pools and runs are different and challenging in their various formats – and alter dependent on the weather and river level. The peace and tranquillity of walking the river and seeing the wildlife - deer, golden eagles and other bird life - is the antidote to the hurly burly of modern life. Good craic with the gillies adding a further critical dimension.

If visiting the area, there are various other activities and places that should be on your radar:

Trout fishing on the multitude of hill lochs scattered throughout the area. Walking from Achfary to Kylesku over the hills provides great views - both North and South. The deserted beaches of Sandwood Bay and Oldmoreshore with their white sands and turquoise water (on the right day) should be on your itinerary. A visit to the bird sanctuary on Handa Island is well worth the short boat trip from Tarbet.

A couple of slightly more off the wall and unusual activities associated with our particular party has been the search for the long lost whisky bottle at Loch Nashelaghan! Another relates to whale snot hunting on Durness beach.

Having visited the area over many years, the hospitality, kindness and friendship shown by all those we've encountered, as well as the companionship of a great many friends, epitomises our Laxford trips. Places and great memories are the panacea to wellbeing. Recalling our visits to the far North during COVID-19 has been a tonic.

Authors & Books
The Kerracher Man – Ian MacLeod

A book providing an insight to the joys, trials and tribulations of living in this remote part of the Highlands.

Musicians & Popular Music
Leonard Cohen – Hallelujah

As a wild child growing up in the era of Cohen - this song holds memories. It has been recorded by many artists in a multitude of formats: listen to the classical arrangement, also a pipe band version - apt ref this week's location.

Composers & Classical Music
Peter Maxwell Davis – Farewell to Stromness

The story behind this piece of music is worth investigating, and the sentiment is relevant and relates to the far North of Scotland and the Islands – with an unexpected and added significance in 2021.

Fine Wine
2000 Vincent Dauvissat Chablis Grand Cru Les Clos, France

A memorable magnum drunk in the company of family and friends at the Laxford in 2011.

WEEK 2
BLINMAN & PARACHILNA AUSTRALIA

Two weeks in, you may begin to recognise a pattern – rugged, wild and remote places appeal.

Blinman fits the bill, and is to be found in Australia's Flinders Ranges. It's not every day you're invited to hitch a ride with good mates in a private plane to explore the Outback. Hence our reason for being there, and I can attest to this being the perfect mode of transport for doing so (if ever you're lucky enough to receive such an invitation).

As mere poms, we've so little understanding of the vast area Australia covers - this being an eye-opener.

We spent a couple of nights on a sheep station with our amazing host 'Ian', who looks after an enormous tract of land on his own, other than calling in help at certain times of the year - such as shearing sheep. UK farmers tend to have a 4 wheel drive and quad-bike at their back door – Ian a small aeroplane. UK farmers talk about how many sheep per acre – Aussies talk about the number of acres per sheep, and at that time suffering a 3rd year of drought.

Sheep farming in the Australian Outback is not for the faint-hearted.

Ian, the perfect host, picked us up from Blinman (Airfield) which doubles up as the local horse-race track a couple of times a year. As you can deduce from my photograph, the facilities can at best be described as sparse. Andrew, our pilot (and friend), was unwilling to land on Ian's own airstrip at the back of the house - which bore very little resemblance to such!

We took a trip out to visit a settlement and mine that had long been abandoned – the local kangaroos not the least interested in us, and otherwise engaged. That evening we went way up into the hills and watched a spectacular sunset and see just how vast the land and plains to the north of the Flinders Ranges are.

We also engaged in a local tradition and downed a few 'Tinnies'.

17

The following day we travelled by Toyota Land Cruiser through the Gorge from Blinman to Parachilna – a rugged and spectacular trip. As you can see from my photograph: the road being on the other side of the creek, Parachilna was closed for the season and we were unable to partake in the kangaroo, camel or emu at the local hostelry.

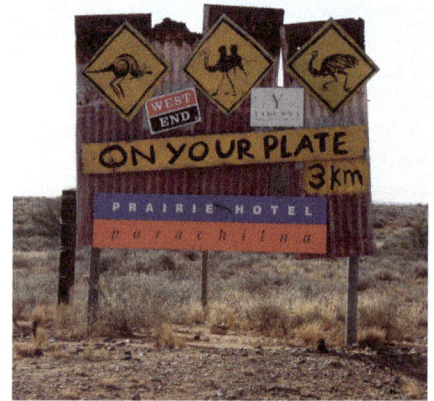

Standing in the middle of the railway line, taking a photograph of the vast wide-open space as it disappears way into the distance, eloquently demonstrates the size and scale of the Australian outback. The main highway was similar and totally deserted, other than a group of 'Emus' diving out of the bushes and across the road immediately in front of us. Obviously waiting for unsuspecting tourists, as we never saw another vehicle all day.

As ever, one of the joys of our visit was meeting 'Ian' and hearing all about his way of life – his hospitality and our lodgings were excellent. The lesson being: What makes trips special and memorable are the people.

You'll hear more of our Outback adventures as the 'Sabbatical' progresses.

Authors & Books
An Outback Life – Mary Groves

A fascinating insight from an early
life in Melbourne to embracing an
'Outback' way of life - a great tale.

Musicians & Popular Music
The Beatles – Long and Winding Road

As a teenager in the '60s with long hair
and a totally irresponsible approach to
life; the Beatles having an influence.
This song has great meaning, and is apt
given our road trip to and from
Parachilna.

Composers & Classical Music
Ludivico Einaudi – Questa Notte

A favourite of Granny and
Grandpa's, and fits well with Ian's
relaxed approach to life at Blinman.
Worth pursuing other compositions
by this amazing artist.

Fine Wine
**2012 Shaw and Smith M3 Chardonnay,
Adelaide Hills, Australia**

Not sure of the wine at Blinman -
so picked a South Australian wine
enjoyed with family at Glen Lyon
(Scotland) on my 61st birthday.

This

week's location invokes a panoply of emotions and memories, and was a pilgrimage my brother and I made to the infamous 2nd World War battle field of Monte Cassino - where our late father fought with the 7th Gurkha Rifles.

To understand how blessed and lucky we are in life requires us to view the sacrifice and suffering of others. Composing this at the time of a global pandemic' makes the events of Monte Cassino highly pertinent.

We flew to Naples and spent a night in the city. A word of warning: having hired a car, we foolishly attempted to drive into the centre of Naples to our hotel. With hindsight, driving in Naples is something to be avoided at all cost.

Suffice to say it was an interesting and hair raising experience – not one to be repeated.

Day 2 and we drove from Naples to Cassino. On the way my brother suggested we stop for lunch at a small town called Caserta, as there was a place of interest marked on the map. Imagine our surprise when we came across the 'Casa Royale' built by the Bourbon Kings in the 1800s and on a scale with Versailles.

This amazing building comprises 1,200 rooms, and has a water feature extending to the far horizon – it's spectacular.

Something we learned on our trip from documents our father compiled shortly before his death: he spent a night in the Casa Royale on route to Greece, having narrowly escaped a court martial due to his truck being stolen outside a Hospital in Rome - while in pursuit of a young nurse. There was more to his story than we knew!

We travelled on to Cassino and met our guide 'Adriana', who chaperoned us and explained much that took place in 1944. It's not my intention to retell the history of the Battle for Monte Cassino - much more astute and capable people having done so. What I would like to convey is an insight as to the enormity and horror of events.

Between January 17th and May 18th 1944, the Allies attempted to break through the Gustav Line – in their path was the Monastery at Monte Cassino. Four major battles took place over this period, and the Allies suffered 55,000 casualties before the Poles eventually liberated the Abbey. The topography has to be seen to be believed.

Anyone who's ever heard the song 'We are the D-Day Dodgers' may comprehend the sour taste of victory - as D-Day totally eclipsed what had been achieved by the Allied forces in taking Monte Cassino. Only latterly did my father speak of his involvement in the battle – and admitted suffering nightmares to his dying day.

The cemeteries demonstrate the massive loss of life on all sides of the conflict - as well as highlighting the diverse range of nationalities and faiths. The heart rending aspect being the ages – many in their teens.

Over a couple of days we visited all the major sites involved in the conflict. Certain stood out, such as Point 593, the front line observation post overlooking the Monastery, where our father served as Battalion Intelligence Officer. The Polish cemetery, war memorial and museum being another.

The most poignant moment was a chance meeting with a charming Italian gentleman at the site of the Battle of Rapido River, which took place from the 20th to 22nd January 1944. The American casualties amounted to 1,330 killed or wounded, and 770 captured over a 3 day period – a major debacle achieving little.

It turned out my new Italian friend grew up in Cassino, and was a child at the time of Rapido River.

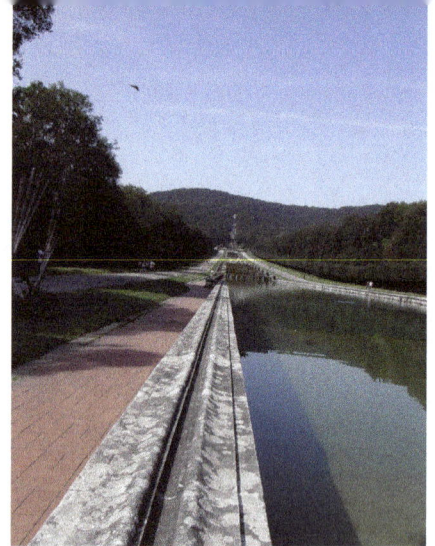

His father was press-ganged to go out after dark and search for dead or wounded German soldiers and drag them back. His mother was a nurse/orderly at the German field hospital. One evening their house suffered a direct hit, and a piece of shrapnel lodged in his head. He ended up in hospital, in a coma, and the German doctor pronounced there was no hope. His mother begged to look after him and in the morning he miraculously regained consciousness and survived to tell his tale.

Henceforth he was known as 'Angel'. He explained the dreadful suffering endured by the local inhabitants both at that time and long after the War. It made our final destination to the restored Abbey and Monastery of Monte Cassino slightly bittersweet, as 'Angel' informed me the amazing restoration took place way in advance of Cassino Town and the welfare of the locals.

Authors & Books
The Monastery – F. Majdalany

A small book I found in my father's possessions after his death; it epitomises the horrors of the Battle for Monte Cassino. A poignant moment reflects on the childlike boots marking the graves of fallen Gurkhas.

Musicians & Popular Music
**The Hollies –
He Ain't Heavy, He's My Brother**

A song from my distant past, but relevant to the pilgrimage with my brother to Monte Cassino - as well as the many 'Brothers in Arms' at this infamous battleground and the 55,000 (Allied) casualties inflicted on their comrades.

Composers & Classical Music
**Aaron Copland –
Fanfare for the Common Man**

An American composer and one I admire, and a fitting piece of music aimed at recognising the unsung and unknown heroes of the 2nd World War.

Fine Wine
**2000 Dom Perignon,
Champagne, France**

Not consumed at Cassino but one Christmas with my sister-in-law in memory of David, my late brother-in-law – a special wine and apt for the memories this week's location invokes.

WEEK 4
SAMOEN
FRANCE

I took up skiing late in life – my 60th year. Better late than never – and something I should have engaged with much sooner. A multitude of bumps, bruises, torn ligaments and a broken bone – I still adore it. What took me so long, I'm not sure: But the joy of nature in spectacular wide open spaces, plus the adrenaline rush from descending the side of a mountain, is pure nectar.

At a time when travel is forbidden and lockdown due to COVID-19 remains in place, the thought of amazing vistas, fresh air, exercise and the camaraderie of mates, – plus an element of après-ski thrown in – is the thing of dreams. It's hard to believe our last trip was early January 2020 - more than 12 months ago. Images, memories and planning a return provides solace.

Mass humanity and crowds are not my bag – as a country boy, I'm more attracted to wild and remote places. Samoen in the Grand Massif region of France is more reserved, quiet and localised than many ski destinations - and for that reason fits the bill for the likes of me. It's also where I learned to ski and first met our legendary ski instructor, character - and friend - Herve.

The Sabbatical is not only about locations, but much about people and memories - and Samoen sits there with the best.

Our original visit to Samoen was via an invite from friends to join them at their chalet in the village. The best piece of advice I can give anyone silly enough to take up skiing late in life - is commission the likes of Herve. He was/is a great friend of our hosts, having taught their family and a multitude of our mates to ski. Engaging Herve was a stroke of genius.

I've never been so black and blue from head to toe as I was that 1st week. Herve's remit was to get me to wherever our party were having lunch and home again. It was an exhilarating experience, and left me totally hooked.

I learned quickly: When Herve pointed out some horrendous slope from the ski-lift and would say "Jamzz, zee likes of you will never do that". The next day I could almost guarantee, it's exactly what I would be doing. A baptism of fire but worth the pain.

Samoen is charming and retains much of its local/regional character, and the inhabitants are extremely friendly. We've a host of restaurants and watering holes our team frequent when in residence - the same being true on the slopes.

The final day of last year's trip was special and memorable – even more so given the benefit of hindsight. We'd no inclination as to what was to befall us within a matter of weeks - all holidays and trips placed on hold due to COVID-19.

It's not every day you have lunch in a restaurant at 3,842 metres above sea level!

The Aiguille du Midi is the nearest to the summit of Mont Blanc the likes of me are ever going to achieve. It's an exhilarating gondola ride to the top, and the views are panoramic and spectacular. We watched some intrepid individuals (as my photograph demonstrates) whose return journey was on skis, first having to traverse the snow ridge on foot – very 'Upton Park'.

We'd an excellent lunch with a fine bottle of wine, as well as a toast and tear, in memory of 'Sarah', a special young lady whose life was sadly and unexpectedly cut short, having just learnt of her death. As a non-religious person, but someone who admires the diverse faiths of others, the Aiguille du Midi (Way up in the Sky) was a fitting and apt place for introspection.

Authors & Books
First Light – Geoffrey Wellum

As a young man, Geoffrey Wellum was a 'Battle of Britain' fighter pilot. Surveying the world from a high vantage point made this book the natural choice.

Musicians & Popular Music
Ray Davies (Kinks)
also Kirsty McCall rendition – Thank You For the Days

Two weeks after Samoen, I was in the car listening to a mum and dad talk about losing their teenage son to 'Sudden Death Syndrome', this song providing them with a semblance of solace. It never fails to make me think of Sarah.

Composers & Classical Music
Claude Debussy – Clair de Lune

Clair de Lune by French composer Claude Debussy is a favourite, with its haunting and dreamlike melody – apt in a variety of ways in relation to this week's location and activities.

Fine Wine
2009 Château Giscours 3eme Cru Classe, Margaux, Bordeaux, France

We splashed out on a Margaux way above Chamonix - to mark a momentous and memorable occasion. I'm not sure of the maker or vintage at our lunch, so I've taken another from an earlier trip to the Laxford (Scotland).

WEEK 5
NEW YORK
USA

Cities are not top of my priority list – other than in short bursts, but then I like to be on my way.

If you've never visited New York, it's worth putting this amazing and unusual place on your list. The scale, height, density and mass humanity crammed into such a constricted piece of land surrounded on all sides by water makes this different and unusual.

Having visited twice, I comprehend its charm and appeal – it's vibrant, alive and very 24/7.

Our 1st visit was not long after the 9/11 disaster and destruction of the Twin Towers. The site had been cleared and was a massive gaping hole in the ground. A moving experience having watched the original TV News footage of this dreadful event on September 11th 2001 - and aware of the enormous loss of life: 2,977 people died.

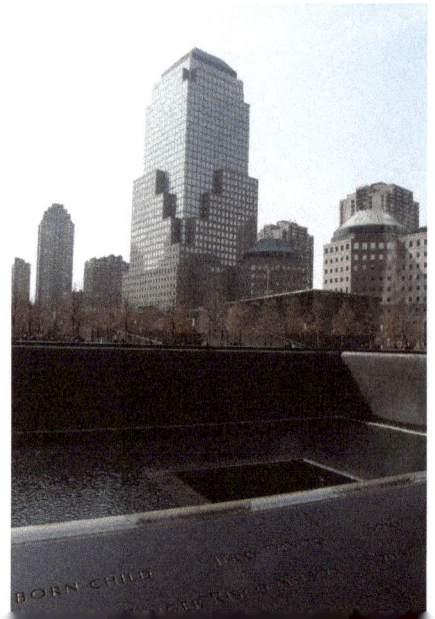

On our 2nd trip a few years later (visiting our son and daughter-in-law, we made our way to the 9/11 Memorial Museum. The enormity and scale of the museum - built into the bowels of the earth - is vast, evocative and fitting as to all that took place on that fateful day. You cannot be other than affected and impressed by this amazing memorial and the telling of the story surrounding this appalling event, and the many individuals involved.

We also paid a visit to the One World Observatory atop the World Trade Centre (Freedom Tower). The views from the 94th floor, at 1,776 feet - are spectacular and panoramic. My pictures hopefully demonstrate - how amazing.

A favourite image of mine is one I took on our first visit. As someone with an aversion to queues, we decided to go to the top of the Rockefeller Centre rather than the Empire State Building. The picture was taken while lying on my back in the street - which aptly demonstrates the scale of the building.

Nothing like this in our home town of Berwick-upon-Tweed!

I enjoy architecture, and Grand Central Station is like a great Cathedral and well worth a visit. Not sure we'd recommend the scallops, as Granny ended up with food poisoning!

Our trips to New York make-out that we are culture buffs, but nothing could be further from the truth - me being a total philistine. But I can recommend a number of other buildings and museums. I love the design and style of the Guggenheim; the Frick Collection is quite something - but it was the building and architecture that did it for me. MoMA (Museum of Modern Art) appeals to my passion in relation to design, and I also enjoyed the Met.

A couple of outstanding aspects of New York (for me) are the High-Line and the Brooklyn Bridge.

The bridge is obviously iconic and features in a myriad of movies and TV series – well worth a walk along the dockside and then across the bridge. The High-Line I love, as it's a walk through a metropolis on a 1.45 mile-long elevated linear park with buildings towering above and created from a former New York Central Railroad spur.

As a country lad and one who enjoys open spaces you cannot go to New York and not visit Central Park. This amazing amenity in the centre of the city covers 842 acres of parkland. The city fathers who enabled Central Park are to be congratulated on providing such an amazing facility for residents and those visiting New York.

Our trips were far from restricted to culture - we enjoyed a variety of restaurants and bars. Other highlights of our New York trips relate to seeing a show on Broadway (Matilda), but the real stand-out one for me being dinner and music at the Birdland Jazz Club - which dates back to 1949, and is an iconic venue.

So if ever you get to New York, you might like to add some of these to your itinerary.

Authors & Books

Tuesdays With Morrie – Mitch Albom

The book relates to a re-kindled relationship between an American University professor and a past student, his mentor courageously facing a terminal illness – much more inspiring than you may think.

Musicians & Popular Music

John Lennon – Imagine

While in Central Park we visited Strawberry Fields and the John Lennon Memorial, making this appropriate - as well as the message contained within the lyrics being the antithesis to 9/11.

Composers & Classical Music

George Gershwin – Rhapsody In Blue

A celebrated American composer. His composition has special meaning in relation to JJ (Our Grandson) – I've a specific piece of music for each of my Grandchildren. JJ's being Rhapsody In Blue.

Fine Wine

2015 Sandhi Chardonnay Santa Rita Hills, Chardonnay, Sonoma County, United States

Fits the bill in a number of ways. We enjoyed this wine in New York with our son and daughter-in-law. But as you may have noticed from my introduction to the Sabbatical, they now have a son (our latest Grandchild): Sandy.

WEEK 6
BERWICK-UPON-TWEED
ENGLAND or SCOTLAND?

Travel

and memories are not restricted to exotic faraway places – they could well be on your doorstep.

Berwick-upon-Tweed is where I was born, and, to be more precise, 5 miles down-river from the village I grew up, and a locality to which we've since returned. I'm a massive fan of my home town: the history, buildings, walls, bridges, alleyways, trails, pier, promenade, river, sea and people are just part of what continue to intrigue me.

There is much to our small town (which does or doesn't belong to England or Scotland). I leave you to check out the history in relation to James the 6th of Scotland and 1st of England regards the 'Borough of Berwick-upon-Tweed'.

Granny and I've been lucky enough in life to travel – but love returning home. We get great pleasure in showing friends from around the world the gem on our doorstep. (We should possibly keep this to ourselves.) Places evoke memories and when you've been involved throughout your life, recording the detail would be a lengthy tome.

Berwick's appeal (to me) lies in its multitude of walks and trails. Nowhere else in the UK has a set of Elizabeth Bastioned walls. In fact, there are few others in the world - Lucca in Italy the best known. How many towns offer the opportunity to walk their circumference on a unique set of ancient walls, with amazing vistas and vantage points?

Berwick-upon-Tweed has another unique trail courtesy of L.S. Lowry, the artist, who came to our town over a period of 40 years and painted many scenes - which can be enjoyed to this day. The Berwick-upon-Tweed Lowry Trail (in which I played a small part) extends to 5 miles and if you wish to see our town at its best, it is well-worth the effort.

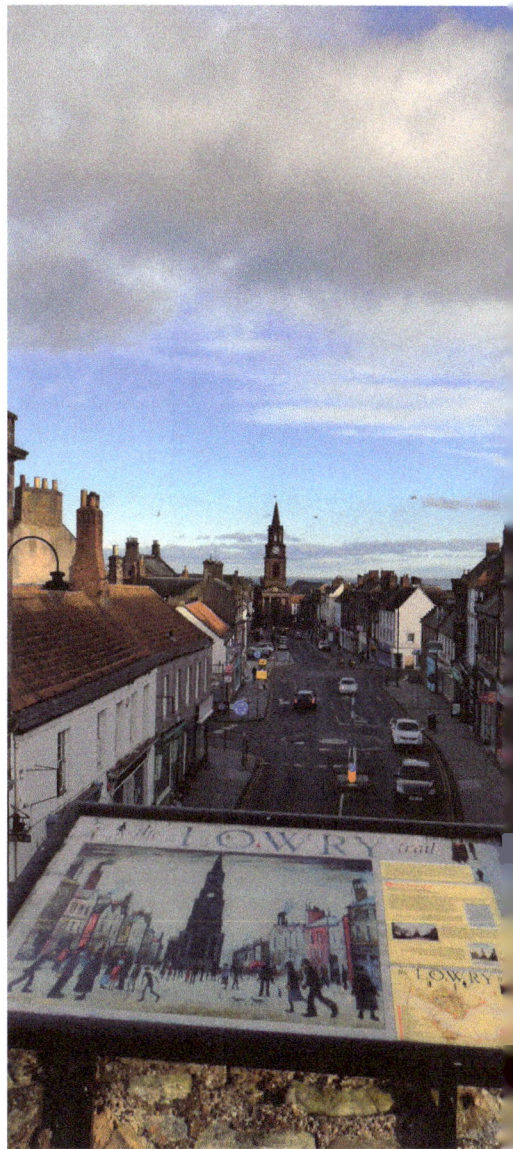

I walk Berwick's walls, alleyways, trails and byways on a regular basis throughout the seasons, and Lowry's paintings have added to my knowledge and understanding of our unique town - also the sea, which straddles its eastern flank.

Berwick's bridges dominate when viewing the town from up, or down, river. They contrast in design, style, age and height. The Berwick Bridge (Old Bridge) built between 1611 and 1621. The Royal Border Bridge (The Railway Bridge) built by Robert Stephenson between 1847 and 1850. The Tweed Bridge (New Bridge) built between 1925 and 1928. My Dad lined the route, as a young schoolboy, when the then Prince of Wales (later King George VI) opened the bridge.

A great way to see all 3 bridges is to walk up-river from the Quayside, along the New Road - which demonstrates and accentuates their scale and contrasts. A detour up-hill to Coronation Park and back down through Castle Vale Park provides panoramic views of the bridges. Before so doing, continue on the New Road briefly and wonder at the amazing driftwood sculpture of a boat created by a talented young local artist.

I've always been intrigued by all the alleyways and passageways throughout our town. To highlight this when back at the quayside, check out the number of ways of going both under the walls, and coming back down from the walls.

Berwick's many walks and trails are on a par with my fascination for the architecture and buildings.

The Town Hall, the Barracks, the Parish Church (built in the time of Cromwell), the Pier - all stand out. But for me, it's all about the individual houses and buildings throughout the town. 50 years ago, Berwick reputedly had the largest percentage of listed buildings at risk in the country! We, the citizens of Berwick, owe a massive debt of gratitude to an enlightened group of individuals who had the foresight to establish one of the first Preservation Trusts.

Check up on the work of the Berwick-upon-Tweed Preservation Trust and the state of our town at that time.

The fascinating aspect of the Trust's history and legacy is their work continues – and of equal importance, the efforts of enlightened residents and local builder's embarking on restorations, renovations and conversions. Nikolaus Pevsner wrote about Berwick in 1957 and described it as "One of the most exciting towns in England". I believe he would be impressed with much that has, and is still, taking place since writing those words.

Authors & Books
The Railway Man – Eric Lomax

Two reasons for including this book. The 1st relates to Eric being a Berwick resident up to the time of his death in 2012. The 2nd: a number of my Father and Mother's friends were (as very young men) Japanese prisoners of war on the infamous Thai/Burma railway. What they endured, and their exploits in later life, are an example to us all.

Musicians & Popular Music
Joe Cocker – With A Little Help From My Friends (Live)

The secret to life and happiness has been family, friends and living in a special place - hence this particular choice. Takes me back to crazier times, and seeing the film version of the 'Woodstock Festival' with Joe Cocker performing.

Composers & Classical Music
Antonio Vivaldi – Four Seasons

A great favourite, having seen Nicola Beneddetti perform live at Malvern. Berwick-upon-Tweed is a location Granny and I see and enjoy throughout all 'Four Seasons' of the year - hence the relevance.

Fine Wine
2008 Château Musar, Bekaa Valley, Lebanon

Consumed locally at Granny's 60th birthday party, which involved a memorable and rowdy wine tasting. Well worth delving into the history of this amazing wine maker – an inspirational story.

WEEK 7
SAILING CROATIA

I first sailed in this glorious part of the world 36 years ago, when it was still Yugoslavia.

It's poignant I did so in the company of my late Father-in-Law, Donald, who died before Christmas of a combination of COVID-19 and dementia. One of your Granny's memories recited at his funeral relates back to that time: -

Dad had a number of holidays without Mum (3 to be precise) – all with my husband! The 1st involved sailing in Yugoslavia – the country still communist at the time. The flotilla was state-controlled, and rules and regulations were scant – a bygone era. They sailed along with an airline pilot.
My husband tells the story of the man from the Navy (Dad) and the airline pilot being in charge of navigation – the issue being, according to them, they were a mile on shore. Hugging the coast became the order of the day. Other escapades such as Dad goose stepping-down the pontoons due to a German boat refusing his kind offer of help. Much hilarity and fun was had by all when in Donald's company. After their 3rd holiday together Mum banned Dad from further solo trips with my husband.

That first trip was memorable in so many ways, and left a lasting desire to return. Over the last 10 years we've been lucky enough on a number of occasions to be invited by great friends; to join them on their boat in Croatia. Being on the move from one location to another, and the scene changing throughout the day, works well for me.

While researching the imagery to use for this week's 'Sabbatical', I came across a thank you letter relating to one of our trips. This, for a variety of reasons, has been censored, but gives an insight to sailing in Croatia:

To the Captain and the Boss (the Lovely Lady Captain),

Thank you once again, and as ever we're in your debt for another magical mystery voyage full of memorable escapades, shenanigans and veritable feasts washed down with copious quantities of wine. Our voyage began in Frapa where both the Captain and Lady Captain were slightly under the weather, which may well have related to the thought of having to put up with us for the next ten days.

Our 1st foray was to the island of Vis where we moored in the bay off Komiza, an exceptional spot. A good time was had by all other than the Captain getting lost on the way home (in the dark) via a trip around the catacombs (in the tender). The following day saw a move to the town of Vis and an unfortunate encounter with Croatia's most inept and notorious marinara and the destruction of our bow thrusters – this being dealt with by our Captain attempting to deposit said marinara in the dock.

The next port of call was to the island of Solta at the marina in Maslinica - which was quite an experience for us mere mortals. Highlights were drinking orange cocktails with the Captain and his entourage, followed by a Venetian masked ball from which we had to escape after the Captain's decision to retaliate and dive bomb our fellow guests! Two days of rest and recuperation followed at the stunning setting of Pipo, where our host's power of authority over ordinary mortals was aptly demonstrated by the local restaurant cooking our food and bringing it back on board for us to eat – Scotch whisky and charm being the required currency.

A night on the island of Palmizana followed with recognition that the average age of our party was nearer three, than two times that of the other customers partaking of the delights of the Drift Wood bar - but the senior citizens proved equal to the challenge. Our last night out on the water was spent at a buoy off the island of Hvar opposite Stari Grad – the boys escaped for a brief sojourn to the Ship Wreck bar prior to another memorable feast served on board. Our final port of call was back in Frapa where we religiously scrubbed and polished the boat under the watchful eye of the Lady Captain, the bow thrusters repaired, and the vessel readied for her next voyage.

Our last day was spent pleasantly anchored off Primosten chilling out prior to our return home and a well-earned (and required) rest. The legendary stamina of the Captain and Lady Captain left us in total awe as ever. So once again a massive thank you for a memorable trip, another to add to the archive.

From your humble Servants and Lowly Crew

I believe this says it all – a magical place, in the company of great friends, and creating lasting memories.

Authors & Books
Eleni – Nicholas Gage

A book given to me by my Father involving a conflict in which he participated – between Communists and Royalists in Greece during the 2nd World War. The story of Tito and the Yugoslav partisans bares some correlation.

Musicians & Popular Music
Katie Melua – The Closest Thing To Crazy

Apt and relevant given this week's 'Sabbatical' location and associated escapades – a musician I much enjoy.

Composers & Classical Music
Camille Saint-Saens – Danse Macabre

A piece of music that neatly encapsulates and represents the pace of life and exploits on our Croatian sailing trips – in particular when related to attending a 'Masked Ball' within a medieval castle!

Fine Wine
2008 Berry Bros. and Rudd Champagne by Mailly, Grand Cru, Brut, Champagne, France

Shared with our great mates (The Captain and The Boss) at a wine tasting in London in appreciation of their generosity and friendship.

WEEK 8
ROSS RIVER
& BIRDSVILLE
AUSTRALIA

Previously I mentioned returning to

our 'Outback Odyssey': After Blinman, we flew to Alice Springs, which must be almost central within Australia's enormous land mass (flying in a small plane makes you realise how vast). It's not every day you see a salt lake – we missed an unusual phenomena by a month, having been full of water and birdlife.

Alice Springs is a major metropolis compared to our other Outback destinations. On arrival in Alice we drove on to Ross River Homestead. Ross River does not live up to its name (certainly not to someone who lives on the banks of the River Tweed) – or at least at the time of our visit. My picture says it all – not a drop of water.

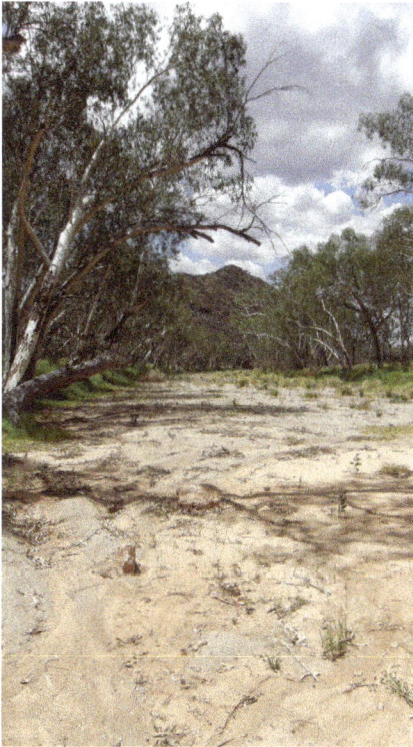

It's amazing, the words of wisdom that emanate from the mouths of babes. We were welcomed by a very excited young boy who'd just lost something precious - but gained something greater in its place. He proudly showed off a cage which the previous evening housed his prize canary. In its place was a python who'd slithered through the bars; having consumed the canary, it could no longer escape. Our young host was happy with the outcome.

On our travels over the years we've come across amazing coincidences and connections back to friends and places elsewhere. Here we are on the other side of the world and very much in the middle of nowhere. Talking to the couple who ran the operation, it turned out he'd previously been the chef at a hotel in Glen Affric (Scotland), a place yet to feature in the 'Sabbatical' – he'd encountered various of our friends in what is a special place.

The following day we went exploring and borrowed a rather dilapidated 4WD. I should first explain the temperature was in excess of 40 degrees – very hot for a couple of fair skinned Poms. Said 4WD was not happy in the heat and boiled the radiator. The decision was made to abandon ship and give it time to cool down, so we set off on foot.

The scenery and rock formations were amazing, but after an hour of trekking Granny started to feel the heat and was no doubt suffering from dehydration. We made our way back to the vehicle to find we had a major dilemma. We could fill the radiator with water, or give it to Granny. The lesson being, when in the Outback make sure you are well-provisioned. After much debate we shared the contents with both parties - and survived to tell the tale.

At the time of our stay at Ross River it could be described as a little tired – hence our hosts made the decision to leave a day earlier than planned. Thus an impromptu and memorable night in Birdsville! Birdsville appears a sleepy little town (general population 115)! An unusual thing, and a 1st for us: the airstrip runs alongside the town and taxied the plane to the front door of the hotel. We also learned about the town's iconic annual event where the population grows exponentially and much fun is had by all – The Birdsville Races.

The 2 day race meeting typically attracts in excess of 7,000 people to the middle of the Simpson Desert – many travel overland or via plane to enjoy the region's yabby races, street party, concert and carnival. I can only image everyone is in motorhomes or under canvas, as there's just the one solitary hotel.

An event that should be added to the bucket list?

I mentioned a memorable night! We appeared to have the place to ourselves until the local roustabouts called in to quench their thirst before going home to eat.

We were challenged to play pool on a table with a definite home advantage. It took Andrew and I until the early hours of the morning to work the table out – or possibly our new found friends were handicapped by the vast quantity of beer consumed by this stage.

We won the game and were high-fiving and dancing round the table when we noticed total silence had descended on the pub. Beating the locals on their home turf was obviously not the done thing. A bit of quick thinking saved our bacon - announcing "the drinks are on us". A memorable night, and major hangover in the morning.

WEEK 9
SADDELL CASTLE SCOTLAND

We've

only visited Saddell Castle on the Mull of Kintyre once and for a few short days. But it's a place I would like to return for a variety of reasons. Our trip related to celebrating my brother's and his mad Doctor mates' 60th birthdays. Our too-brief visit left a variety of lasting memories - which is what the Sabbatical aims to achieve.

Saddell possibly demonstrates a bit of an obsessive nature - as you may deduce from this week's pictures! My academic career never took off and came to a premature end. I met a lovely lady a few years ago who possibly had the answer. For the past 40 years I've been involved in the design and manufacture of kitchen furniture. Hence the reason our paths crossed at an open day for one of our dealerships near Oxford.

She wanted to understand the process involved in creating her 'Dream Kitchen'. To cut a long story short, I explained when designing I see things (in my head) via 3D and colour – a flat plan computes as the finished article. She promptly informed me I was dyslectic. The previous week she'd been talking on the subject at the O2 in London to an international audience. I've never followed it up, but believe – whatever it is – for me it's a blessing, not an issue.

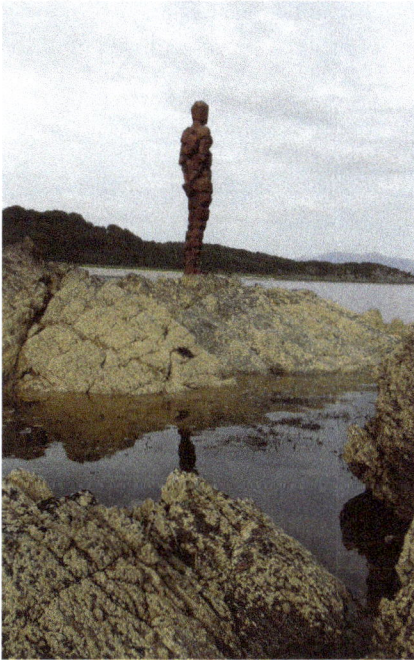

The point I'm attempting to make relates to my relationship with imagery and how I use this to conjure up great memories, and in a time of COVID-19 and lockdowns that's good for one's mental health. I'm getting a bit philosophical so back to Saddell Castle, the Mull of Kintyre and a memorable trip. The ingredients being a beautiful location, great weather, family, friends, fun, laughter, food, wine, imagery, and Anthony Gormley!

Prior to our visit I'd no inkling that Sir Anthony Gormley had placed one of his amazing statues on the rocks in front of the castle. With a Geordie mother and wife, I'm a fan of Newcastle-upon-Tyne and adore the Angel of the North. So as you can see, I became slightly obsessed with his sculpture and how it changed due to time, tide and weather.

I should explain we did other things besides rushing up and down from the statue whenever the sun was out, or the state of the tide. My brother's family and friends are much more cultured and talented than I; hence we were entertained with musical soirees, singing, a play, and Scottish country dancing. A great time was had by all – the mixing of the generations without inhibitions being a long-time family trait, and one I highly commend to others.

A further reason for wishing to return to Saddle is the drive. We didn't have the luxury of time on our side, so I've no photographic evidence – but take it from me, the views and scenery were spectacular. Another time!

As you may have previously gathered, I've a fascination with design and architecture. The restoration of Saddell Castle by the Landmark Trust deserves recognition and praise. They took on the project at a time when it would have been deemed beyond redemption by lesser mortals, and the finished article is a sympathetic homage to its history.

One particular design feature from its distant past relates to entering the front door of the Castle: – If you were deemed unsavoury or an enemy - you plunged through a removable floor to the dungeon (without an exit) below. I should emphasise the Land Mark Trust have made the Castle much more homely and inviting.

This is a charitable organisation well worth supporting and they've many more interesting properties within their portfolio.

To my brother, sister-in-law and joint hosts – a massive thank you for a memorable event and visit.

WEEK 10
SINGAPORE

As I've previously alluded to, cities are not high on my priority list. But we decided to break our journey and as we'd never visited Singapore, made the decision that would be our stop-over destination. We arrived very late due to a delay, and I was decidedly grumpy. The next morning was baking hot and the humidity stifling. Granny decided we should explore - but we were defeated and ended up back by the pool on the roof of our hotel.

Having recently been around the Australian Outback, as well as an amazing trip to Stewart Island / Rakinura off the Southern tip of New Zealand, mass humanity was possibly my issue. To put things in perspective, Stewart Island (which we'll return to in a later Sabbatical submission) is more than twice the landmass of Singapore – Singapore has a population of 5.7 million in comparison to Stewart Island's 402! I think my comfort zone was being tested.

Singapore's history and place in the world deserves respect – but a hot and travel-weary Brit was not in the frame of mind to explore and to take this in. A beer and a swim were more on the cards.

The scale of the economic miracle Singapore has wrought from a small landmass was easy to comprehend from the vast amount of shipping we could survey from our vantage point. Singapore has always intrigued me in relation to the Japanese offensive in 1942 and the subsequent surrender of the Allied forces. The Northumberland Fusiliers (of which my grandfather was a much-decorated 1st World War officer) having arrived just prior to this.

I mention this as a number of my Father and Mother's friends became Japanese prisoners of war on the infamous Thai/Burma railway after Singapore's surrender. I'd grown up knowing and respecting a number of these. It had been our intention to visit the Changi Museum, where many thousands were interned at the notorious Changi Jail. As explained we never made it - something I regret as it would have been a fitting homage to Mum and Dad's friends.

At the time of a global pandemic it does no harm to try and comprehend and understand the suffering, trials and tribulations of others – it helps put things in perspective.

One fact I learnt on our visit, of which previously I had no inkling: Singapore and Holland have something in common. The country's territory is composed of one main island, 63 satellite islands and islets, and one outlying islet. The comparison relates to Singapore's landmass having grown by 25% since the country's independence in 1959, as a result of extensive land reclamation. I'm often accused of being a mine of useless information!

The most enlightening thing came about after dark and totally altered our perspective on Singapore. We went up town for dinner and were bowled over as to the City coming alive, humming and vibrant. The variety of restaurants and food vendors made a decision on our dinner destination difficult – but as you can see from the imagery it was buzzing and full of life. I should add the food was delicious.

After dinner we walked about and took in the views, of which I hope my photos provide a notion as to what this entailed. It's fascinating and vibrant - the architecture very much in the same vein. We ended up with a completely different viewpoint and perspective from the start of our day. As they say never judge a book by its cover.

Authors & Books
The Forgotten Highlander – Alistair Urquhart

I read Alistair Urquhart's obituary and had to read his book; a young soldier in Singapore at the time of the Japanese invasion. Subsequently on the Death Railway, followed by the Hell Ships to Japan, torpedoed and one of very few survivors, forced labour in a mine at Nagasaki, and witnessed the atomic bomb being dropped!

Musicians & Popular Music
Otis Redding – Sitting On The Dock of The Bay

Apt given our morning looking out over the bay and later dinner at the dockside.

Composers & Classical Music
Arvo Part – Speigel im Speigel (Cello and Piano)

A sombre but uplifting piece that fits the bill given this week's thoughts and sentiment. I first heard this at the time our Granddaughter (Lozzie) was born, and always associate this with her.

Fine Wine
2014 Domaine Vincent and Sophie Morey, Chassagne-Montrachet 1er Cru Les Embazees, Burgundy, France

Not a clue what we drank in Singapore – so a wine from a special occasion drunk with friends in London.

WEEK 11
CHEVIOT HILLS ENGLISH/ SCOTTISH BORDER

This week marks the anniversary of the first UK COVID-19 lockdown, which deserves mention. The vaccination programme is progressing well, with over 50% of the UK adult population receiving their first dose – Granny and I included. Hopefully there's light at the end of the tunnel - but unfortunately the EU is heading for a **third** lockdown. So foreign travel is not yet an option - but hopefully the ability for UK trips is about to become a possibility?

If this proves the case, Granny and I've plans to visit various locations we've experienced and loved at different times in our lives. The Sabbatical was never restricted to exotic and faraway places – we'd also like to demonstrate to you (our grandchildren) what's on our doorstep and a little further afield. This week's destination a great example.

We see the Cheviot Hills from our house in Northumberland, which evokes happy memories going back a great many years. But I'll start with a more recent and unusual trip. My birthday present from Granny a couple of years back was a flight in a glider from Millfield Aerodrome at the foot of the Cheviots. A great way to see the world – from a great height and the background noise of the wind. A brilliant experience, and one I highly recommend.

My picture shows us (I had a pilot) being towed way up high by a small plane and suddenly you're released and at the whim of the wind and thermals – it's exhilarating and the views spectacular. The Sabbatical aims to conjure up memories of good times and places we've loved – I should also add experiences such as flying in a glider!

Hidden in the Cheviots is an amazing history going back thousands of years - as my image of a Bronze or Iron Age hill fort confirms. My knowledge is limited, so I'm looking forward to a project currently under development in Wooler - relating back to the time of 'AD Gefrin' (1,000 years ago) when this was the most powerful kingdom in the land!

The 'Ad Gefrin' project involves a distillery and visitor centre to tell the tale – a worthy addition to the Sabbatical?

My history in relation to the Cheviots is much more recent and goes back some 60 plus years. My Mum took us as children to 'Happy Valley' where we would build dams, swim, play, picnic – very simple pastimes, but ones that stick in the memory bank. In turn we took our kids – the next generation. We also spent memorable times in the 'College Valley' where great family friends had a shack.

The most remarkable thing given the world of today: As children (very early teens and below), we would cycle the 20 miles from our village to the 'College Valley' (without adults and no gears on our bikes) and spend the week running wild. Facilities were sparse: no electricity or running water, and an outside chemical loo which was not for the faint hearted. Basic, but we had a ball out on the hills, in the stream guddling fish (now highly illegal). Memorable!

A more recent trip (4 years ago) to 'College Valley' involved great friends hosting the most amazing wedding for their daughter. The church service took place in Norham Parish Church (Cromwell's headquarters at one point) – a beautiful service – and then bused to the reception in a marquee in the 'College Valley'. A stunning location, as my picture demonstrates.

A fitting and lasting memory of a lady who lived with cancer for 11 years – a model to us all in extracting every last positive ounce out of everything she ever engaged with.

This week's Sabbatical entry has been rather inward looking and philosophical due to attending the funeral (last week) of a good friend who helped Granny in the garden. Fit and healthy up to Christmas and now no longer with us - due to cancer. Watching a 12 year old boy bury his mother is hard to compute. It made me dwell on my own father and great friend who lost his beloved mother (my grandmother) at the same age.

A gentle reminder to those following the 'Sabbatical' – a small donation to Cancer Research UK could make a difference to the lives of others.

www.cancerresearchuk.org

Authors & Books
The Isolation Shepherd – Iain R Thomson

The book, as the title alludes, involves life in a remote location – very apt in relation to the Cheviots. Possibly should accompany another Sabbatical submission (Affric and Glen Strathfarrar), but I've another book earmarked.

Musicians & Popular Music
REM – Everybody Hurts Sometime

REM's song inevitably conjures up memories of a special couple (good friends and colleagues from my industry) and played at Mark's funeral after he lost his inspiring fight with a brain tumour.

Composers & Classical Music
Ralph Vaughan Williams – The Lark Ascending

Fits the bill soaring above the Cheviots and the sentiments evoked in this week's Sabbatical entry.

Fine Wine
2016 Domaine Hubert Lamy, St Aubin 1er Cru, En Remilly, Burgundy, France

The 29th March sees the 1st opportunity in a while to sit down in the garden with my sister (relaxation of COVID-19 rules) who has inspired me over the last 2 years. Her dignity, stoicism, attitude and approach to dealing with throat cancer is exemplary and to be applauded (she is her mother's daughter). A special wine for a special person.

WEEK 12
LAKE MUNGO
& TIBOOBURRA
AUSTRALIA

This week's Sabbatical recounts the last leg of our epic 'Aussie Outback Odyssey' – very special and memorable, and a world away from the current global pandemic restrictions. Lake Mungo a fitting final destination to the trip. This image conjures up great memories: the last remnants of the original homestead were the chimney and the 'Dunny' – I've a picture of Granny perched on the throne; it's more than my life's worth to include this!

Our flight path from Alice Springs to Mungo required a pit-stop in Tibooburra to replenish our fuel. Tibooburra promotes itself as the 'Hottest Outback Town in New South Wales' – we can concur with this. The temperature once the sun went down remained in the 30s. The town is miniscule compared to our village in Northumberland. It's reminiscent of a Wild West cowboy scene – tumbleweed, but unfortunately no show from John Wayne!

We learned a couple of things relating to Tibooburra's past: - A Brit (Charles Sturt) led an expedition in 1845 to find Australia's 'Inland Sea' – they dragged a 27ft long whaleboat all the way out into the desert, eventually abandoning this in the vicinity of Tibooburra. On the outskirts of the town is a memorial and full-size replica of the boat.

My picture shows the Family Hotel; we paid a visit for a sundowner in their unique bar. We learned two things: first, the raunchy murals on the walls where painted by Clifton Pugh around 1969 – when he was marooned by flooding (The Big Wet). To pass the time he created his eye catching mural. Second, and significant as Australia is currently suffering once in a century floods, Tibooburra has a couple of months warning prior to these engulfing them.

Should you ever choose to visit the Australian outback, Lake Mungo and the Walls of China should be on your itinerary. The story told by the visitor centre gives an insight to the amazing history of the indigenous Aboriginal people who inhabited these lands many thousands of years before the white-fella ever appeared. The oldest human remains in Australia (Mungo Man) were discovered here in 1974 - and date back some 40,000 years.

Lake Mungo is situated in Mungo National Park, a World Heritage Site. I don't intend to explain the detail, as much more able and qualified people have done so. Mungo comes across as a magical and spiritual place – the vast landscape and silence makes it ethereal even for someone like me with no religious leaning. It makes you contemplate and want to learn more of the indigenous people who led such a fascinating and sustainable existence.

We, the people of today, can learn much from their way of life.

Before returning to Mungo Lodge we stayed and watched the sun go down on the desert and Walls of China – special, memorable and thought-provoking. Our accommodation at Mungo Lodge was brilliant and the culinary fare served up by the French couple running the facility was excellent, plus we had the first decent wine we'd encountered on our 'Outback Odyssey'. We were extremely well looked after, along with the one other couple staying.

Mungo proved a fitting final destination to a once in a lifetime adventure. As my picture demonstrates, this included a multitude of firsts, such as the plane parked alongside our accommodation (being far from normal in our lives). Another was taxiing along the runway to chase off the kangaroos before attempting take-off.

To our great Aussie mates and Andrew for piloting and guiding us on an amazing 'Odyssey', a massive 'Thank You'.

EYEMOUTH & ST ABBS
SCOTLAND

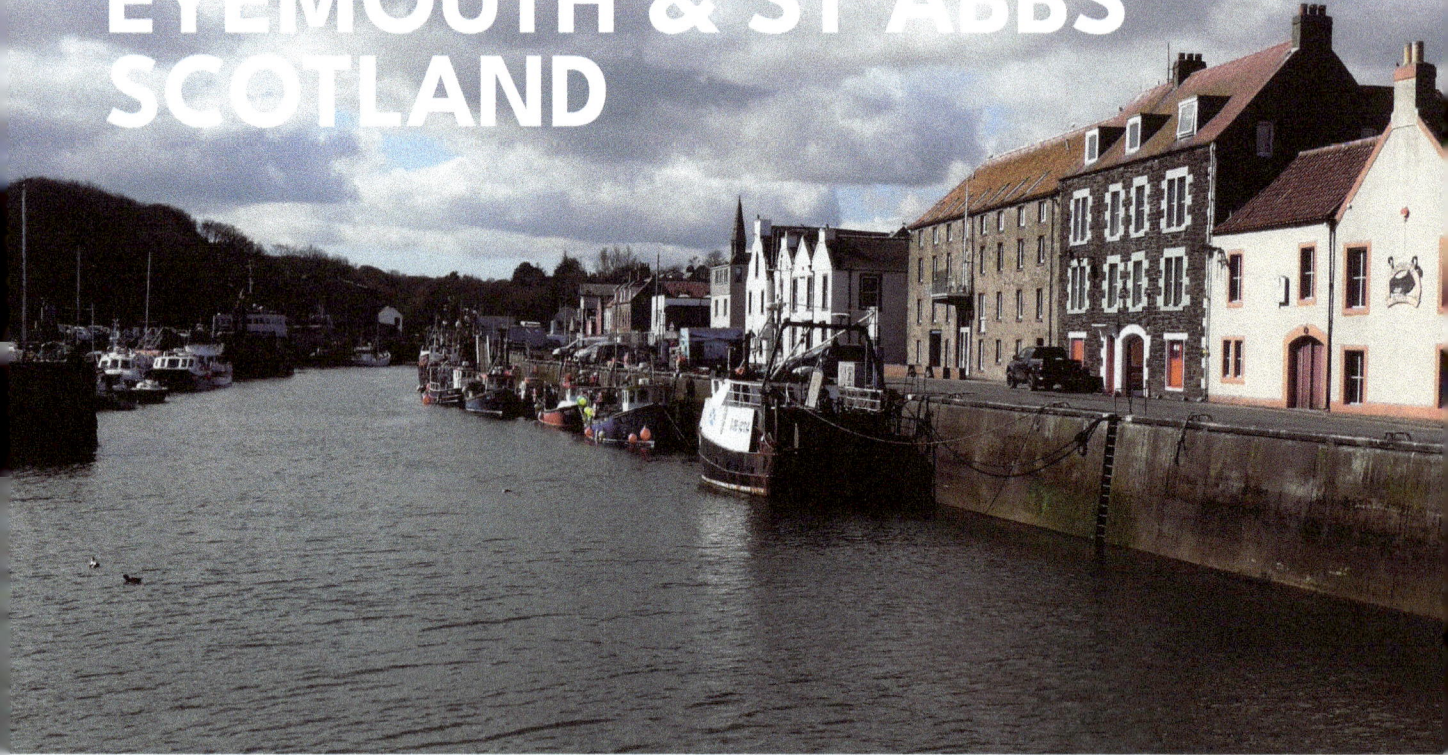

It was never our intention 'The Sabbatical' should concentrate on faraway exotic locations – the inclusion of places we've been lucky enough to visit and enjoy in the UK an important consideration. The pandemic and lockdowns over this last year have restricted our visits to a great many of these, but reiterates how fortunate we are to live in the English/Scottish borders and all that's on our doorstep.

It is pertinent that elections for Scotland's devolved parliament will be held in a month's time. My Father saw himself as a Scot, my Mother English, and I was born in Berwick-upon-Tweed (which for many years belonged to neither). I'm a proud Borderer and Northumbrian – a foot in both camps. Living on the English side of the Tweed and looking across the river to Scotland, at no point in my life has ever appeared an issue.

I'm hopeful today's journey (one we've made over many decades) just a few short miles up the coast to Eyemouth and St Abbs remains open to us in the future and, as previously, we're made welcome.

Eyemouth is a place I've visited as a child, with our children and grandchildren. At a time of COVID-19 and the tragedy this has inflicted on so many families, Eyemouth has its own dark tale to tell. In 1881 the town and adjacent fishing villages lost 189 of their menfolk – 'The Eyemouth Fishing Disaster' (Black Friday). A devastating storm led to the loss of much of the local fishing fleet.

As my first picture highlights, Eyemouth harbour remains a thriving fishing port. The pilgrimage destination when visiting over many years has always been Giacopazzi's on the quayside for one of their famous ice creams, or - dependent on the time of year - fish and chips. Eyemouth has a fascinating history as told by Gunsgreen House, a fine mansion overlooking the harbour, built on the proceeds of smuggling.

A few short miles up the coast is the pretty village of Coldingham which has an amazing beach and bay with its many multi-coloured beach huts – not a sight one expects to see on Berwickshire's coast.

The contrast with the next step of our day trip could not be more diverse. The rugged harbour and village of St Abbs flanked by high rocky outcrops and cliffs. Once again this remains an active fishing port, but also a thriving location for recreational divers.

An interesting fact relating to St Abbs: in 2015 the RNLI (Royal National Lifeboat Association) withdrew the village's longstanding lifeboat. The locals objected and raised funds to buy their own, and now run this independently. Tunnocks Teacakes played a major role - so inevitably we have one with our coffee at the café adjacent to the harbour.

St Abbs Head National Nature Reserve immediately north of the village is a wildlife haven, and one of our long-time favourite walks. It's a 4 mile circuit and takes in spectacular views, and incorporates a lighthouse warning shipping of the wild and dangerous coast they are navigating. We often walk the trail in reverse, which provides the most spectacular views of St Abbs harbour from our vantage point way up on the cliffs.

In relation to the sea, it would be remiss not to mention the death this week of Prince Philip (the Duke of Edinburgh). A man who epitomised an old-fashioned approach and attitude to duty and responsibility. I grew up around many such people, involved at a very young age in fighting a War. I'm in awe of the amazing lives they went on to lead. The Duke's life and achievements are well worth examining.

An interesting point in relation to my Uncle - your Great, Great Uncle. I'm led to believe he and Prince Philip were both young officer cadets at Dartmouth Naval Academy, prior to the outbreak of 2nd World War.

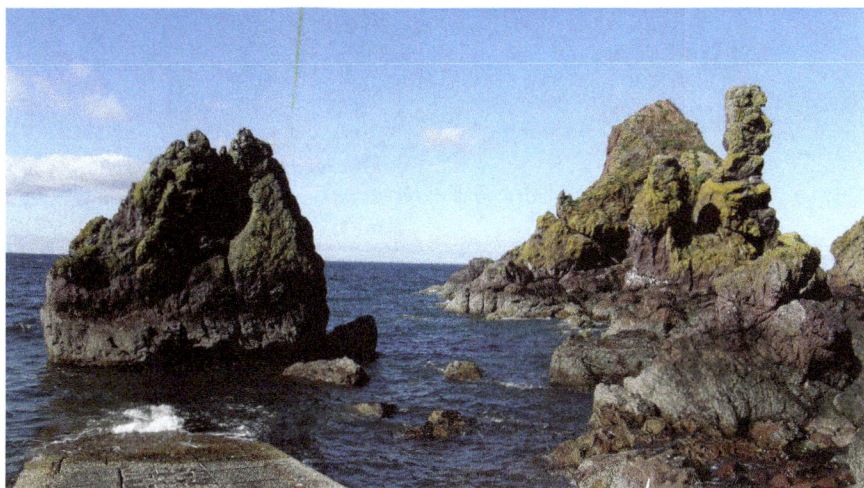

Authors & Books
The Mauritanian - Mohamedou Ould Slahi

The importance of free speech, civil society, law and order, a willingness to look at and engage with opposing sides of an argument, facing one's own prejudices and ignorance, learning to change your point of view. A book which makes you appreciate the value of such: 14 years imprisonment and torture - without charge!

Musicians & Popular Music
Joni Mitchell – Both Sides Now

A lady whose music takes me back to a different time in my life - the late 1960s and early 70s. The ability to look at life from both sides (in my experience) makes for much more interesting people.

Composers & Classical Music
Erik Satie – Gnossienne No 1-6

A thought-provoking composition given the varied topics and subjects incorporated into this week's 'Sabbatical' submission. One I relate to when attempting to rationalise complex issues.

Fine Wine
2005 R. Lopez de Heredia, Vina Tondonia Reserve, Rioja, Spain

As we're local this week, the wine relates to the last Christmas (2019) we managed to congregate together as a family, in the Borders. The COVID-19 pandemic curtailed this year's festivities – hopefully next Christmas?

WEEK 14
CORVARA
ITALIAN DOLOMITES

It's

been a sad sight to watch the webcam footage of Corvara in the Italian Dolomites – a place we've been lucky enough to visit on a number of occasions. As you can deduce from my picture, it's a magical destination for skiing and scenery - and one Granny and I **ha**ve much enjoyed. Normally the place is buzzing and alive with people. Unfortunately COVID-19 and the latest lockdown in Italy appears to have curtailed resumption of any sense of normality.

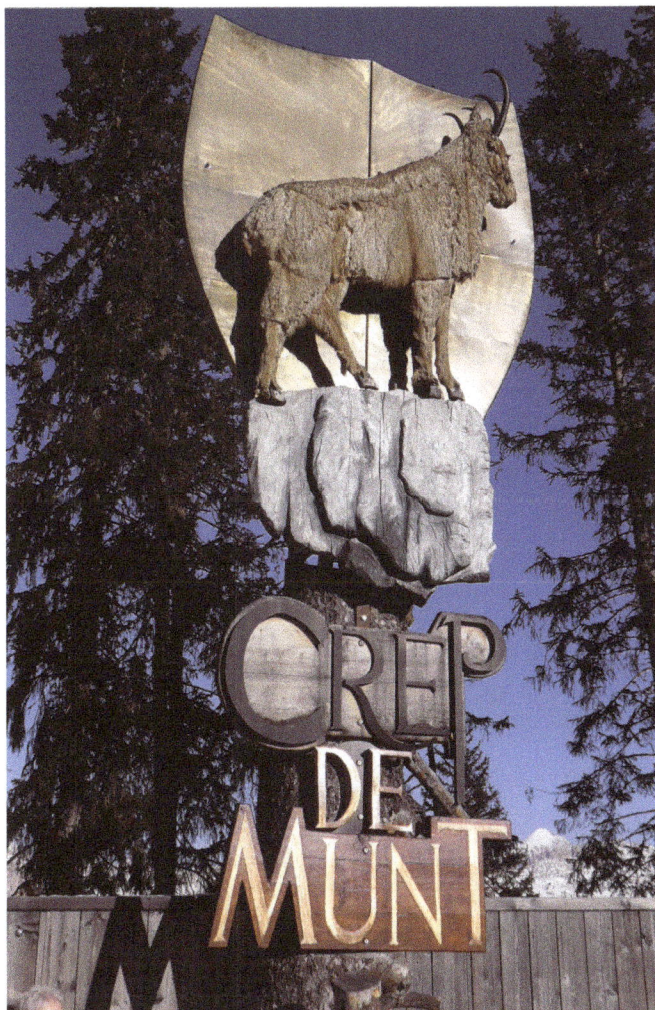

Our thoughts are very much with our Ladin friends and hope they're all in good health, and next season (with luck) we may be in a position to pay a return visit. From our various trips we've learnt Ladin is the language of the Alta Badia and South Tyrol region. They see themselves as Ladin rather than Italian or Austrian – similar in ways to us (English/Scottish) Border folk. Our experience being they (as us) are extremely proud, warm and hospitable people.

2020 and the COVID-19 pandemic has made Granny and I realise just how lucky we are to live where we do, and our business activities have not been curtailed to the extent of so many others. The economic consequences of running a business in a ski resort, unable to operate, puts one's own issues into perspective.

We thought it appropriate to mention a few of our favourite watering holes when in Corvara. The crew running the Hotel Table have indulged our party on each of our visits. A special mention of Lara and Walter, whose bar and service is beyond reproach. Tommy and team up on the slopes at Crep de Munt deserve mention. L'Murin at the bottom of the slopes is a place we've been known to frequent and quench our thirst after a hard day's skiing!

I can report the breadth of skiing in the Alta Badia region is exceptional, with two stand out opportunities available:

First, the 'Sella Ronda', described as one of the most famous ski carousels in the world. Sella is a mountain in the Dolomites from which 4 valleys lead. The valleys are joined by road passes but, in winter, over 200 lifts and 500km of slopes provide an alternative method of travel which allows one to ski all around the mountain.

Secondly, the 'Hidden Valley' involves a spectacular gondola ride to the top of the mountain and then you're off on a one way journey with no option other than to get to the other end. Plagiarising someone else's words "There are few ski slopes that offer a true feeling of isolation, wilderness and solitude". The Hidden Valley provides all of these.

From prior Sabbatical submissions, you may deduce I'm intrigued by much that relates to the two World Wars - the Alta Badia region played a major role in the 1st. Many people believe the 1st World War was only fought in Flanders, but that is a grave misapprehension.

Before I explain more: the 'Don't Mention the War Game' requires to be aired. My erstwhile friends decided anyone uttering the word 'War' had to pay a penance with a circuit of the hotel, out in the snow. This was aimed at me and set-up in advance. This left no alternative other than to go on the attack - after 3 days, they were suing for 'Peace'!

On our trip to the Hidden Valley and gondola ride up the mountain we encountered numerous gun emplacements and fortifications dug into the cliff face. The battle for the Italian Front or Alpine Front took place between Austria–Hungary and Italy from 1915 to 1918. The conditions must have involved unimaginable horrors – the topography, logistics, conditions, snow, and ice must have made survival, never mind fighting, an almost impossible task.

I would like to finish by explaining that trips with our ski buddies are always courteous, civil and with much fun and hilarity had by all. An oft-used phrase being: if you take yourself too seriously, or unable to laugh at one's self, you're with the wrong crew. Granny and I recognise and subscribe to that analogy.

So to all our great mates: when the opportunity arises, we'll be there – that free ski pass is coming into sight!

Authors & Books
**Captain Corelli's Mandolin –
Louis de Bernieres**

A book given to me by my late Dad, and one that provides an insight to an aspect of Italy's involvement in the 2nd World War. A lovely tale with a tragic twist and end result – with a truth that should be told.

Musicians & Popular Music
**Procol Harem –
A Whiter Shade of Pale**

Takes me back a very long way in time and the colour aptly relates to snow – seemed a worthy choice for this week.

Composers & Classical Music
Wagner – Ride of the Valkyries

Evocative and representative of my style of skiing (or total lack of), too often on the edge with inevitable consequences. This piece of music epitomises what often leads to my downfall.

Fine Wine
**2013 Cascina Luisin,
Barbaresco Rabaja, Piedmont, Italy**

To all our skiing buddies: I've a case of this in my cellar, **and** plan and look forward to the day we can all get together to discuss and arrange our next ski trip. I will provide this fine Italian wine to celebrate.

WEEK 15
CHRISTCHURCH
& DOME HILLS
NEW ZEALAND

GRANNY and I've paid one brief but amazing

visit to New Zealand (The South Island) and would love to return. Unfortunately with the current **global pandemic**, New Zealand is closed to the likes of us - so **it'**ll have to wait.

We flew into Christchurch from Tasmania with great mates from the Borders. A major bonus and font of insider knowledge; our party comprised a Kiwi now living on the other side of the world, not far from us. Our 10 day road trip was an absolute riot of catching up with her family and friends - as well as being royally entertained. As you will learn, we didn't stay in one place for long, and the legendary Kiwi hospitality is all it's made out to be.

I've deliberately not included pictures of Christchurch, as it suffered a major earthquake 12 months prior to our visit and evidence of this was all around. 185 people lost their lives in the disaster. It makes you count your blessings when you encounter houses and areas totally destroyed and uninhabitable. The word 'liquefaction' was added to our vocabulary: a major impact on the devastation which meant various areas can never be rebuilt.

We encountered a minor after-shock on our first night, but were well anaesthetised due to the local bonhomie.

The following day involved a boat trip up the Waimak (Waimakariri) river in the jet boat of our Kiwi pal's brother David. Brilliant fun, and highly recommended if ever the chance arises. Their father was a fighter pilot during the 2nd World War. After hostilities he and a group of mates, searching for an adrenaline rush, engaged with these amazing boats which come hurtling out the water over the shingle banks and back in again.

We'd a great day; the picture demonstrates racing home ahead of the weather.

I've previously mentioned coincidences relating back to our part of the world. We were invited to a 3 generation family gathering. I sat beside Uncle David, who was engaged in writing a book collating their family history. A year or so previously it transpired he'd paid a visit to a house adjacent to our village (in the Borders) in which one of his ancestors (Mr Alder) had lived. It turned out we were related in some convoluted fashion.

The next leg of our journey involved a drive to Dome Hills in Otago. A beautiful setting way up in the hills – a total contrast to Christchurch. Our hosts Cindy and David, as all the Kiwis we encountered, could not have been more amenable and hospitable. Cindy had, in an earlier life, spent time in the Borders and knew many of our friends.

We'd a couple of memorable escapades while at Dome Hills.

The first related to a trip around their amazing spread of land to see the new dairy operation and learn about Kiwi sheep farming. This involved a Toyota Land Cruiser with a mighty engine that had (in the circumstances) to work hard for its keep. Our trip took us way up into the hills and, as my picture shows, into the clouds. We got slightly lost and ended up coming down the other side of the mountain – all off-road.

This provided the opportunity to stop at a historic watering hole the 'Danseys Pass Inn'. Wouldn't like to do the reverse journey after a heavy session in the pub! Sufficiently refreshed, we came back via Danseys Pass; a spectacular unmade road wending its way through the mountains – my picture highlights the views.

The 2nd memorable trip involved David's jet boat on Lake Benmore. Interestingly, Benmore Estate (Scotland) will make an appearance in the 'Sabbatical'. Lake Benmore is big, and we picnicked on the far shore. Once all there, we had a thrash about in the boat – only to find it was filling with water from a broken pipe. We made shore, but the car and trailer were a 17 mile hike around the lake. Luckily we concocted a repair, utilising one of the ladies' lipsticks.

To Cindy and David, a massive thank you for all your hospitality and generosity – we had a ball.

Authors & Books
Ship of Fools – Fintan O'Toole

An apt title following this week's escapades. This small but thought provoking book was presented to me by a charming young Irish girl in Spain (a head on her shoulders way beyond her years), who recounted the devastating consequences the Irish financial crisis wrought on her family.

Musicians & Popular Music
Enya – Evening Falls

To the days of my Sony Walkman. A haunting piece of music that takes me back to times past and great friends (no longer with us) - fits the bill in relation to the possibility of being marooned on the shores of Lake Benmore.

Composers & Classical Music
**Karl Jenkins –
Concerto grosso for strings 'Palladio' 1**

Flying over the water or haring down the sides of mountains in beautiful surroundings – this composition nails it.

Fine Wine
2015 Craggy Range, Aroha, Te Muna, Pinot Noir, Martinborough, New Zealand

Not a scooby as to what we drank at the time – so a fitting New Zealand wine drunk in the company of my great friend Goram and my No.1 Son at Café Murano (London) in 2019.

71

GLEN AFFRIC
& FASNAKYLE
SCOTLAND

My love affair with the Scottish Highlights came about via fishing on the Laxford in Sutherland 30+ years ago. Followed shortly after by Granny and I's introduction to Glen Affric via great friends inviting us and children to join them and family at their Bothy not far from the Fasnakyle hydro power station.

Many memorable trips and escapades have subsequently taken place over the intervening years.

Coincidences and tales to unrelated friends are becoming an aspect of the Sabbatical. Pals from the Borders, Sue and Denbigh (sadly no longer with us); Granny and I were witnesses at their wedding, and we had a number of memorable trips to Aiguablava in Spain. As a child, Sue lived in Cannich while her Dad (John) was a Civil Engineer working on Fasnakyle's Hydro scheme.

Sadly the Bothy at Fasnakyle has been sold. Our last trip involved celebrating my 60th birthday with all the family in tow. I'm left with a crooked finger from that particular trip.

The previous week, while fishing on the Whiteadder – a great day's fishing – unfortunately I slipped and caught my finger between 2 rocks and dislocated it. In such circumstances, I'm told, the best thing is to immediately put it back! Possibly getting out the river first would have been a good idea, as I next came to in the water, having lost a brand new rod and reel, luckily found down-river. I tried unsuccessfully once more to un-dislocate my finger and again passed out – this time on land – at which point I decided to make my way back to the car and take myself to hospital.

My wonky finger relates to the fishing being so good on the Glass and the Farrar on my birthday week - I dispensed with the splint the hospital so kindly provided. The picture of the frozen cobweb was taken at the time, and reminds me of our first night prior to the family arriving. No-one had been to the Bothy for weeks and it was perishing – Granny had so many layers on, she looked like the bag-lady in Alan Bennett's film 'Lady in the Van'.

I must move on, as not only did we have many memorable holidays at the Bothy, but also stayed with a group of mates at Affric Lodge on a number of occasions. Affric Lodge sits at the end of Loch Affric and is one of the most spectacular settings ever. Waking up in the morning with that view from your bed is very, very special.

As with so many places featured in the Sabbatical, it's the people Granny and I've got to know on our travels that makes a place stand out. Affric and Fasnakyle do not disappoint. One family in particular deserves recognition - the MacLennans. Old Duncan, his brother the Blue Charm, and sons John and Duncan made our stays memorable. John the stalker and young Duncan the gillie. Too many tales to tell but I will recount a couple.

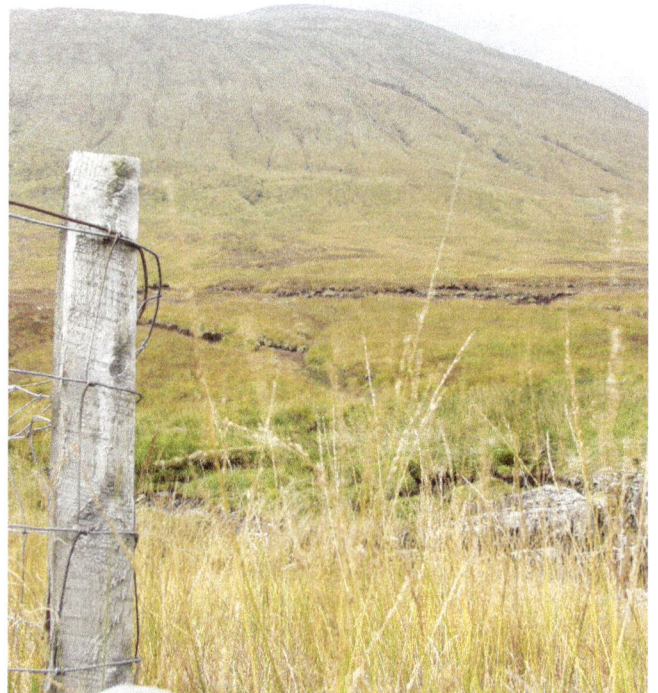

Granny and I were honoured to be invited to a dinner to celebrate John's 60th birthday and 40th year of stalking Affric estate. I, a fisherman rather than a stalker, was asked by John to accompany him on the hill the next day. An invitation I could not refuse. Sitting out on the mountain side discussing all that goes on in John's office was very special and memorable. The ongoing debate as to our contrasting views on metal flies was on the agenda!

Our trips to Affric and Fasnakyle are legend and imprinted on our memory – too many tales to tell, and not enough space. Our visits tended to coincide with my birthday, with many celebrated at the Tommich Hotel. A gentleman a decade my senior shared the same birthday – I wonder if their fishing party still makes an annual pilgrimage.

Another memorable occasion involved an invite to the Stalkers Ball – I should point out to the uninitiated, this relates to deer stalking. Not sure why only 4 of us were staying at the Bothy, but we went like lambs to the slaughter. The Stalkers Ball took place at the end of the season at the Slaters Arms in Cannich. Best description: wild.

Affric Lodge and Estate changed hands a number of years ago and has been gentrified and turned into a prestige luxury upmarket holiday destination – so unfortunately we've not been back. But Affric, Fasnakyle and Glen Strathfarrar are etched in our memories. The scenery and colours later in the year, as you can see, are spectacular.

Special, special memories and thanks to our great mates who first introduced us all those years ago.

Authors & Books
**My Yester Years In Glen Affric –
Duncan MacLennan**

Conservationists and environmentalists could learn from the life of a very special gentleman, who lived his life in Glen Affric for over 100 years. Head Stalker at Glen Affric from 1942 – 1989. An insight to country life and ways that resonate with me.

Musicians & Popular Music
Glen Campbell – Gentle On My Mind

This one's for the Laird, and the significance will not be missed on all involved in the many memorable trips to Affric and Fasnakyle. The title is extremely apt for a magical location.

Composers & Classical Music
Johann Pachelbel – Canon in D Minor

A piece of music Granny and I treasure and fitting in relation to the landscape as well as escapades that will remain imprinted on our memories for ever.

Fine Wine
2012 Mas de Daumas Gassac, Rouge, St Guilhem-le-Desert Cite d'Aniane, Languedoc, France

No record of what we drank on my 60th birthday at Fasnakyle. So a wine drunk in the company of the pair who introduced us to Affric and Fasnakyle - on a fishing trip to the Laxford. Our son the fine wine specialist was mortified when Janet pointed out to him one of the bottles was corked.

WEEK 17
GLEN LYON & WANAKA NEW ZEALAND

Last year's family gathering in Glen Lyon was postponed due to the impact of the Pandemic and the re-run is now in our diary for July – fingers crossed. We should point out this will not take place in New Zealand, but Perthshire (Scotland) - a place Granny and I have visited on multiple occasions over the last quarter of a century.

Our love affair with Glen Lyon (Perthshire) has a rival, as the second leg of our whistle-stop road trip of New Zealand's South Island, involved visiting an alternative Glen Lyon. Both amazing and magical in their own truly individual ways. But we'll reserve our explanation of the Scottish version for July's Sabbatical submission.

Glen Lyon (NZ) is hidden away in the mountains, and happens to be the home of our Kiwi mate's cousin. Useful having someone in the know. The entrance to their homestead passes an amazing lake, as you can see from my picture, and it only gets better the further you progress onto their spread. Mountains, rivers and glaciers – you will note these are in the plural. It's a spectacular and awe-inspiring landscape, and a place we'd love to return.

Our hosts for the day, Ken and Jane, took us for a picnic way up the valley beyond the junction of the two rivers. Their favourite spot was denoted by a large boulder known as the 'Dog Rock' – not hard to see how it gained its name.

We've previously mentioned the point: meeting amazing people, in wondrous surroundings and hearing all about their way of life, broadens one's horizons and understanding of the world. People who live in remote wild places inevitably have a different and interesting perspective to us more cosmopolitan souls and Glen Lyon (NZ) was no exception.

Glen Lyon Station covers 100,000 acres of rugged land in New Zealand's Southern Alps which Ken and Jane have farmed for a number of decades, cattle and sheep being their main activity. It was fascinating to hear about the mustering of the animals – much of this done on horseback due to the nature of the terrain. We were shown pictures of a prior cattle drive; it appears an amazing experience. Fording the rivers in a flood looked an interesting ordeal.

As a fisherman, the stories of the size and strength of the rainbow and brown trout would on its own tempt me back. We learned the wildlife includes tahr, chamois, as well as red and fallow deer – we did look out for these, but unfortunately in the short space of time afforded to our visit we were unlucky. We'll have to make a return trip!

We were left with the impression looking after the land rather than just making money was high on their agenda; a thought and approach to life we would like to pass onto our grandchildren. Glen Lyon Station and its custodians made a lasting impression on us and we'd like to thank Ken and Jane for their hospitality – special.

The next leg of our trip involved a night in Wanaka with our Kiwi companion's Aunt Sally. New Zealand is a bit like the Borders - everyone is related or connected. We were late in leaving Glen Lyon, which involved a mad dash down the road. We blotted our copybook by not arriving on time for dinner with Sally, her sister, and cousin Antonia (very late). But as per the Kiwi way this was not an issue, and a riotous night took place with all sorts of reminiscences.

You'll see this week's composer and music is Nigel Hess and the theme tune from the film 'Ladies in Lavender'. Whenever I hear this composition it brings to mind a memorable evening in Wanaka with Sally, Annabel and Antonina. There was much mirth, hilarity and high spirits - age should never be a barrier to such.

A year on from the devastating earthquake there was much talk of the impact; we learned many people chose to relocate to Wanaka. Its situation in the Southern Alps with lakes and mountains is special. One day is not enough to do it justice - hence a return visit is on the cards. We also learned about 'Warbirds Over Wanaka' – our Kiwi companion's Dad flew Mustangs against Japan during the 2nd World War, this event a tribute to such people.

To all our new Kiwi mates – a big thank you.

Authors & Books
Becoming – Michelle Obama

A book we believe our Grandchildren would benefit from reading at some point in their lives. We can't think of a better place to hunker down and engross yourself in this insightful book than Glen Lyon (NZ).

Musicians & Popular Music
Sinead O'Connor – Nothing Compares 2 U

A piece of music on Granny and I's favourites list and, after 40+ years together, the sentiment is still in play. Could get us into trouble but the title of this week's entry also relates to how we feel about Glen Lyon - which one we cannot possibly admit to!

Composers & Classical Music
Nigel Hess – Ladies In Lavender

A piece of music Granny and I value, and fitting in relation to the landscape as well as escapades that will remain imprinted on our memories for ever. Sally and Annabel epitomise the stoicism and wit of that generation.

Fine Wine
2015 Neudorf Moutere Pinot Noir, Nelson, New Zealand

Funny we don't appear to remember the wines drunk on our amazing whistle-stop tour of New Zealand's South Island – something to do with their renowned hospitality? A white Burgundy is Granny's preferred tipple, but she's partial to a New Zealand Pinot Noir. Hence this week's choice from a memorable wine tasting in London in 2018.

WEEK 18
FORD & ETAL
NORTHUMBERLAND

Week 18 involves a couple of important
milestones: The 1st, Granny and I received our second AstraZenca vaccines
last week and no ill effects. 2nd, we're a third of the way through the
'Sabbatical'. Also of major significance, UK COVID-19 data and relaxation
of Government restrictions appear to be heading in the right direction.

A 12 minute drive from our front door is the village of Etal, a few miles further on is Ford. Both go back to the early days of our marriage, as we lived in an adjacent village (Crookham). We were penniless and the only way onto the housing ladder was to purchase a dilapidated school and carry out much of the work ourselves. This being the catalyst to a career change - property development and renovation, subsequently design and furniture manufacture.

Ford and Etal played an important part in our early married life – a very happy time. Ford and Etal Estate provide our community with an amazing and diverse array of benefits. We're lucky and privileged to have this on our doorstep.

My first image shows Ford Castle, an imposing Grade 1 listed building which dominates the village skyline. This dates back to 1278 with a fascinating history involving events such as the Battle of Flodden. The Castle is to embark on a new chapter in its history - with 2 open-air operas taking place this summer. A spectacular venue and setting.

Ford has history ref our Aussie mates too. The Old Dairy and Antiques Centre (Ford) is a favourite coffee stop for Granny and I, and we also partake in their renowned 'Pop-Up Suppers'. The first ever was instigated by us when looking for a venue to entertain the Aussies. We hit on the idea of asking the Old Diary to put on a special event - the rest is history.

Some semblance of normality hopefully on the horizon?

Which made the pair of us stop and reflect how lucky we've been compared to so many others since the Pandemic hit our shores. We've a detached house and garden, no home schooling, maintained jobs and income, live in a rural location, we've been healthy – to name just a few. Realising and appreciating what's on our doorstep a further subject discussed. This week's 'Sabbatical' submission aims to highlight the point.

81

Travel by road from Ford to Etal and you come across Heatherslaw which 40+ years ago was part of our commute to work. This is no longer possible, as the bridge is now pedestrianised. Heatherslaw has a spectacular, fully-functioning flour mill driven by water from the river Till. A pretty little community, and well worth a visit.

Heatherslaw features an unusual mode of transport - a miniature railway which travels alongside the river to Etal. A trip we've done with the Grandchildren on numerous occasions. A great attraction, but rather fraught when eldest grandson decides to throw his brand new sword out the carriage. Retrieved later by the kind train driver.

Travel by road or walk to Etal and you come across an idyllic picture postcard cricket pitch and pavilion. This has history; my brother-in-law was involved in Allan Lamb's testimonial match – held at Etal. The likes of David Gower, Dean Jones, Lance Cairns, Franklyn Stephenson and other well-known world cricket stars participated. A wild occasion with much frivolity and shenanigans thrown into the mix. It's also the pitch on which our son learnt to play.

Etal, with its Manor House and Chapel, the imposing ruin of Etal Castle, and - as my picture demonstrates - a model village with a thatched pub (the Black Bull). Many years ago while renovating our first home I was visiting Abercrombie's the local joiners and undertakers when the pub went up in flames. A spectacular blaze.

We should point out the Black Bull was recently refurbished and ideally placed for a beer after your walk.

One of our favourite walks is to follow the path down river from the village. This being something Granny and I've done for many years, and gain as much enjoyment as the very first time we came across it. Returning to the point we're lucky as to all we have on our doorstep. We'd like to salute James (Lord Joicey) and the Ford and Etal Estates Team – you've made lockdowns much easier for Granny and I.
Thank you.

Authors & Books
**Where Memories Go –
Sally Magnusson**

This week involved pondering memories and times passed by. Sally Magnusson recalls the highs and lows of her mother's story in relation to dementia. Relevant as Granny's dad (Donald, your great grandfather) had dementia and sadly died from COVID late last year.

Musicians & Popular Music
**Sandy Denny (Fairport Convention) –
Who Knows Where the Time Goes.
Also Eva Cassidy rendition.**

One particular song stood out in relation to this week's sentiments – it evokes contemplation and the passage of time. Takes me way back to my younger days - before Granny's good influence came to bear.

Composers & Classical Music
Samuel Barber – Adagio for Strings

As a young penniless newly-married couple, this time in our lives played a major part in shaping our destiny. Samuel Barber's Adagio for Strings: a fitting and inspiring composition in relation to both our past and present.

Fine Wine
1996 Château Langoa Barton 3eme Cru Classe, Saint-Julien, Bordeaux, France

Seems apt to go back in time to a family Christmas at home in the Borders (2006). We certainly didn't drink such wines in our early days. But with our son a fine wines specialist, our palettes have developed over the years.

83

An invitation to visit the last inhabited island before Antarctica and go 'Whale Snot Hunting' is not an everyday occurrence. This came about a number of years back when Granny and I met a delightful New Zealander (Janet) while on holiday in Spain. The best bit being assurance 'Whale Snot Hunting' would pay for our world tour!

So on our odyssey to the other side of the World we took Janet up on her intriguing offer.

To get to Stewart Island (Rakiura) required driving to Bluff, the southern-most tip of New Zealand's South Island. We broke our journey with a stop-over in Queenstown - another spectacular place in this amazing country. Well worth putting this one on your bucket list – great scenery, hiking, skiing and so much more.

We were alerted to Queenstown's cosmopolitan and social attractions when coming over the pass – in the valley below a major pop concert was in progress. Unfortunately we'd no tickets, so unable to attend. Although we did meet those performing back at our hotel – the Doobie Brothers, a nice bunch of guys.

My first picture demonstrates the view from our balcony looking out over the lake to Queenstown. A fitting venue to host a dinner party for our Kiwi mate's brother and family. We'd an excellent night and brother (Pete) regaled us with his many exploits as skipper of a boat in Milford Sound (NZ). Another place to add to our list.

Day 2 involved a hike out to Sam Summer's Hut (we pondered whether he featured in Anne Summers' family tree)? Unfortunately the accommodation was unavailable for rent – which pleased Granny. We learned about the gold miners and were in awe of their exploits tunnelling a water channel through a vast depth of solid rock.

The following day we made our way to Bluff and got the ferry across to Oban, the only major settlement on Stewart Island's 650 square mile landmass. The Maori name for the island is Rakiura – meaning the Land of the Glowing Sun. An interesting thing happened on the crossing: the ferry stopped to let us observe a Royal Albatross, a great sight.

The island's population at the last census in 2018 amounted to 408 people. Amazing to think both Singapore and Hong Kong, with their populations of millions, would both fit into Stewart Island (Rakiura) with room to spare. Almost 90% of the island is dedicated to National Park. It's a very special place and one we'd love to return to and learn more.

On that first day we had to work for our supper. Pete, Janet's husband, took us fishing – we were successful but had to compete with the Mollymawks, which are related to the albatross family. They were extremely adept at stealing your catch when trying to land a fish. I'd great fun trying to capture them on camera in full flight – very hit and miss.

I can't remember the species of fish, but the meal Janet cooked that evening was stupendous. We'd a riotous dinner with Pete and Janet telling us all about their life running a sheep station – now operated by their offspring. At the time of our visit, Pete was running the helicopter service on Stewart Island. The conversation travelled around the world with many hilarious tales – the one of the donkey in the Middle East is etched on my brain!

Day 2 and we went trekking: as you can see from the picture, the vegetation is dense as some of our party were to learn the hard way. Too much talking and not enough observation meant certain members of the group failed to make that important right turn, which led to a cross country excursion and multiple scratches and lacerations.

Granny and I's foray to New Zealand was drawing to end, as we were heading off on our Australian outback odyssey - leaving our great mates in the hands of Janet and Pete. Our whistle-stop tour of New Zealand left us with only one wish: someday we'd love to return. It was very special and memorable and we cannot thank our buddies from the Borders, their relatives and friends enough for the most amazing trip and legendary Kiwi hospitality.

My final picture taken from the balcony of the house on the morning of departure – brings memories flooding back.

The one question I haven't yet answered: did we find the 'Whale Snot' (Ambergris)? Sadly the answer is no, but our education on this subject may have led to discovering the mother-lode on Durness Beach (Scotland)!

Authors & Books

The Island – Victoria Hislop

At a time of a Global Pandemic it's worth celebrating the amazing efforts of our scientists. The Island relates to a disease (leprosy) which now has a cure - but previously devastated the lives of those afflicted.

Musicians & Popular Music

The Doobie Brothers – Listen to the Music

Listen to the Music is an aspect and aim of the 'Sabbatical' and makes this song a good choice. Meeting the Doobies in Queenstown adds to its relevance.

Composers & Classical Music

Michael Nyman – The Heart Asks Pleasure First

The music from 'The Piano', a powerful and celebrated movie set in New Zealand, makes this the perfect choice. Listen to the guitar as well as the piano rendition.

Fine Wine

2015 Felton Road, Bannockburn Pinot Noir, Central Otago New Zealand

Total amnesia as to the wines we drank in Queenstown and Stewart Island - I wonder why? This is a wine we drank with family at Christmas 2018, as well as with great mates in Melbourne in 2019.

WEEK 20
PAXTON HOUSE & CHAIN BRIDGE ENGLAND/ SCOTLAND

Takes

me back to my earliest days and a multitude of memories. We've been lucky enough to walk out our front door and venture down river on magical walks throughout the Lockdowns. Slightly curtailed in the 2nd of which you will learn more. 1 ½ miles downstream is the Chain Bridge (Union Bridge) – straddles the English/Scottish Border.

This world-class icon of the Industrial Revolution was the 1st of its kind to carry vehicles and celebrated its 200th anniversary in 2020. From a very young child the walk to the Chain Bridge has been an aspect of my life. Great family friends lived in the Boathouse just further on from the bridge - our 2 families inseparable while growing up.

As much of my childhood was spent at the Boathouse as my own home. We ran wild and feral, the river playing a major role in our lives - boats, canoes, swimming and fishing. Roaming the countryside causing mischief was part and parcel of ones development in those days. The bridge a giant climbing frame – totally frowned upon in today's world. Often out from morning to night - returning to the ring of the bell which signified food on the table.

My oldest friend in life took on the house from his parents. This meant Granny and our No.1 Son and No.1 Daughter (your Mums and Dads) also experienced the Boathouse. Far too many events to recount: birthdays, weddings, christenings, anniversaries, lunches, dinners, memorable parties and sadly wakes. A lifetime of fond memories. They've since moved on to pastures new, but we're lucky enough to still visit as good friends purchased the house.

I mentioned our ability to cross the bridge/ border was curtailed in the second lockdown. My picture highlights the issue – the bridge, currently undergoing a complete restoration, has been dismantled. It's been a fascinating feat of engineering, and we look forward to raising a glass at its re-opening in a year's time.

It would be remiss not to mention the work of the Friends of the Chain Bridge and Northumberland and Berwickshire County Councils for making this project possible. They deserve recognition for all their valiant efforts.

During the first Lockdown we regularly crossed the bridge and walked to Paxton House. The weather at times was stupendous and often we'd the place to ourselves. As my picture demonstrates, Paxton House is an amazing Neo-Palladium Adam Mansion and incorporates an outstanding collection of Chippendale furniture. The house and grounds were generously placed in the hands of the Paxton House Trust by John Home-Robertson in 1988.

When I was a youngster Paxton House was very much a private dwelling - the boundary wall somewhere, beyond which we seldom ventured. An early experience of the House came about in my teens when good friends rented a flat in the stable block – a number of lively parties taking place. Since that time the house and grounds have been opened up to the public and we've become Friends of the House and supporters of the Trust's work.

Paxton House is a great venue for weddings, events and concerts. We've attended a great many over the years. Two in particular stand out:

The first was my 50th birthday party in the picture gallery which houses artworks from the National Gallery of Scotland. A memorable night. The second relates to taking the Aussies and others to Paxton Pub - or so everyone thought. In fact, we'd arranged a dinner in Paxton House's palatial dining room. Our convoy of cars arrived at the impressive front steps where Jim (our caterer), dressed as the butler, served champagne; fond memories.

Our annual membership of Paxton House provides us with the opportunity to walk the grounds throughout the year. Hopefully my pictures demonstrate having this facility on our doorstep as and when we wish is an absolute bargain, and one we take full advantage of, a walk along the river and through the woods with the dog a way to destress.

My picture of the cottage, which we refer to as the Gingerbread House, with its sculpture of a long lost pet dog on the chimney stack - a favourite. Another is the Boat-House down by the river – mustn't forget the play park. A more recent addition is the fully operational water wheel and its history. The Edinburgh Window is the latest attraction.

One last tip: on the walk home (once over the bridge), call in for coffee at the Chain Bridge Honey Farm.

Authors & Books

Wojtek the Bear – Aileen Orr

Wojtek spent part of his life a mile or so from the Chain Bridge at Winfield Aerodrome and loved nothing better than a beer, a fag and a swim in the river. Fighting alongside his Polish comrades at Monte Cassino, a tale worth reading.

Musicians & Popular Music

Louis Armstrong – What a Wonderful World

A homage to my parents and their great buddies (Eric and Joyce) this song could not be more apt – amazing times.

Composers & Classical Music

Gregorio Allegri – Miserere Mei

So many memories over a great many years and sadly many of those involved no longer with us – makes this piece of music an ideal choice.

Fine Wine

2010 Jean-Paul and Benoit Droin, Chablis Premier Cru, Montee de Tonnerre, Burgundy, France

No record or memory of what we drank at my 50th birthday or the one-off dinner with the Aussies in Paxton House's elegant dining room. Next best thing: a wine drunk a decade later with family at my 60th birthday.

Granny

and I were recently asked how our involvement and engagement with Australia came about. This week's Sabbatical destination and trip played an important part in taking us to the other side of the World.

It's a long story and goes back to the 1960s and the £10.00 Poms. Family folklore has it my Uncle (Philip), who fought behind the lines in Burma with Wingate's Chindits in World War 2 became so disillusioned with the UK political situation, he threw his medals into the muck-midden and emigrated to Australia with my aunt and 3 cousins.

David, my cousin, married an Aussie maths teacher (Penny) and on their back-pack tour of Europe spent time at our family home in the Borders. David was killed in a tragic accident in the Australian Outback many years ago. We rather lost touch until an invite from Penny to the premier of a friend's musical (Crusade) at the Edinburgh Fringe.

Subsequent to this we met up at various times and occasions when Penny was in the UK – a memorable one being a further invitation to attend another first night of the Impresario's musicals (Eurobeat) in London's West End. Sitting in the front row of the dress circle, as guests of honour, with Cameron Macintosh behind us was memorable.

WEEK 21
COMO & VERONA ITALY

To cut a long story short, our trip came about via a further invitation from Penny and husband (Chris) to join them and a group of their Aussie mates at an amazing villa on the shores of Lake Como. Verona came into the equation in that Granny and I had a few days in this great city before joining up with our hosts and friends.

It's a spectacular place, as demonstrated by the Coliseum (Arena) which at the time was hosting Verdi's opera extravaganza Aida. We much enjoyed Verona and walked for miles, and visited a plethora of sites and antiquities.

A sculpture making a lasting impression was, funnily enough, a contemporary piece in the main square. Our Italian being zero meant a number of years passed before learning its story. Later to become a temporary exhibit on the 4th plinth in Trafalgar Square – and viewed by some as controversial. In Verona we saw a beautiful, inspiring and thought-provoking statue. The story of Alison Lapper (Pregnant) and sculptor Mark Quinn are worth examining.

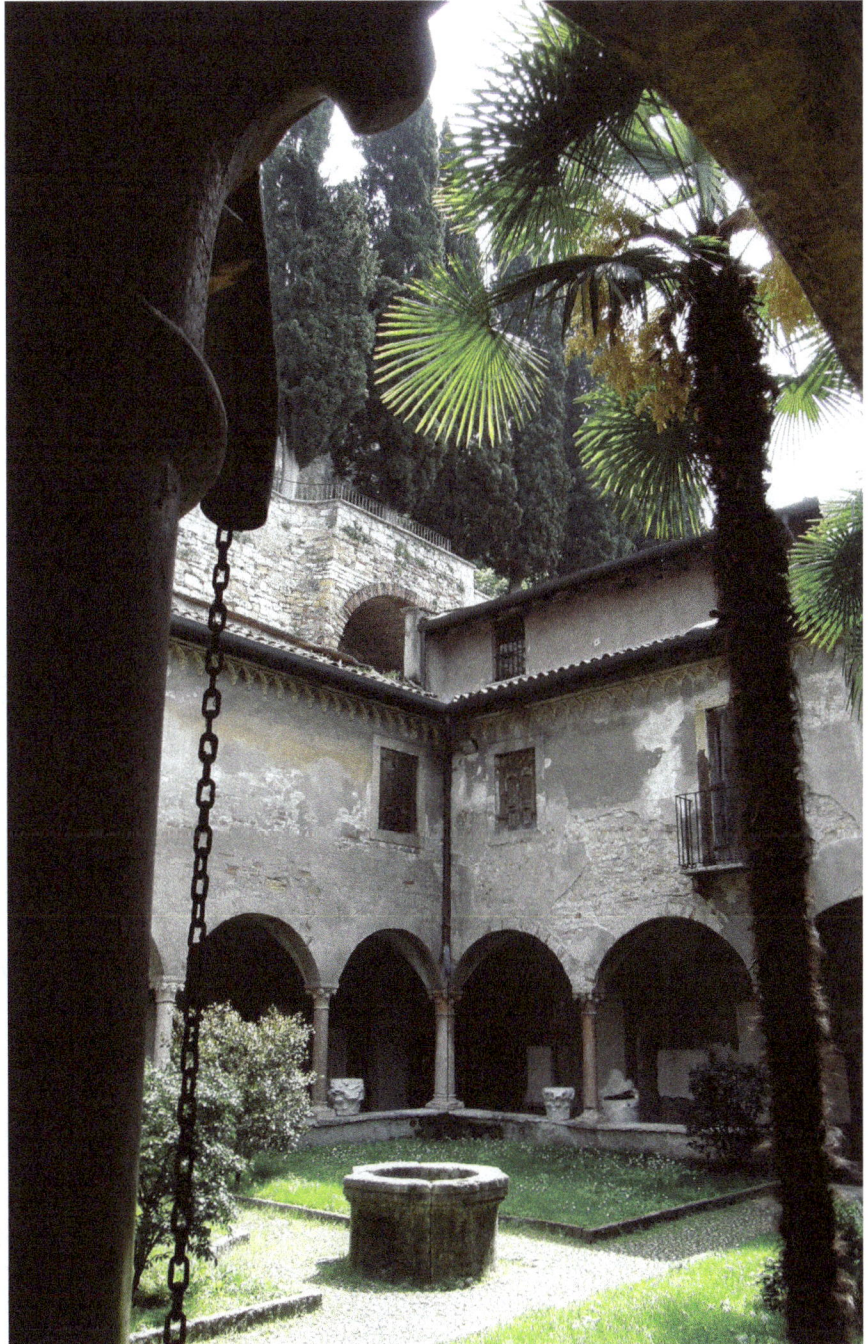

93

Having a diminutive car (a Fiat 500) was a blessing when arriving at Como, due to the roads being extremely narrow and steep. Meeting the local bus head-on, driven by a mad Italian, was a hair-raising experience. The lake and surrounding mountains are spectacular. Sadly our trip to Como goes back a number of years and unfortunately the weather did not appear to shine for the camera – hence my photographic record of our memorable trip is limited.

This week's opening picture demonstrates our amazing accommodation. Accessing it proved a conundrum. We located our destination, but with only one small (pedestrian access) doorway, working out where to park took time. Eventually we twigged you parked in the house opposite and access was via a tunnel under the road.

More bizarre was the red carpet leading all the way to the front door. Turns out the villa had been utilised for a high-end Milan fashion house and photo shoot the previous week, and the carpet still in situ. It turned out we were hobnobbing in good company: a short distance to the left was George Clooney's residence and along the road to the right was Villa d'Este, hotel to the rich and famous, also featured in numerous films including a Bond movie.

My earlier picture looking out over the lake from our terrace and the contented ducks demonstrates how relaxed the situation proved to be. We ate most meals on the terrace, and the scene at night proved equally breath-taking. It turned out the Impresario and his wife were talented chefs – the food stupendous.

Too many tales to tell, but a magical week visiting the likes of Bellagio encompassed a memorable boat trip. The place making the biggest impression was found after a hike in the hills on the opposite shore. Turned out there's a hidden world way up in the trees. Bar Italia in Torno, a village at the side of the lake, became the go-to watering hole.

To answer the original question: we must have passed muster having been adopted by our hosts Aussie mates and invited to visit them back in Oz. The rest is history, and has been a special journey of friendship and learning.

Authors & Books
My Left Foot – Christy Brown

One of 17 surviving children from a brood of 22 and born with severe cerebral palsy. The story of Christy Brown's struggles and challenges, as well as his amazing achievements resonate with those of Allison Lapper.

Musicians & Popular Music
Dire Straits – Romeo and Juliet

A mental picture of Granny and I at the purported Romeo and Juliet balcony in Verona – hence a good choice. Dire Straits features in our story – Granny dancing to Sultans of Swing on the table of a Gullet in Turkey!

Composers & Classical Music
Beethoven – Fur Elise Bagatelle No 25

A piece of music which fits the Aussies laidback attitude and approach in spectacular surroundings – the changes in tempo mirror memories and escapades of a special time and place.

Fine Wine
2011 Pieropan, Soave Classico La Rocca, Veneto, Italy

The combined knowledge of our party in relation to Italian wines was severely limited. So found a regional wine consumed in the company of family and friends on my 60th birthday at Glen Affric and Fasnakyle.

95

Newcastle-upon-Tyne

has played a major role in both Granny's and my life – not least it's where we first met. Being strictly correct, it was outside the city boundary at a pub called the Highlander where she spied me and made the fortunate decision (for yours truly) to come and introduce herself – no social media or mobile phones in those days!

The Angel of the North is a much-loved symbol for us both. Whenever travelling North by road or rail, it signifies we're back home in (God's Country) the North East. Antony Gormley's amazing sculpture also symbolises a bygone era: coal, steel and shipbuilding were massive regional industries in our youth – a very different landscape to today.

Gateshead Millennium Bridge

Granny and family are of Geordie stock, as is my Mother's family; both our children born in Newcastle (Geordie blood in all you grandchildren). Granny attended a Newcastle convent school, she and I two peas in a pod with sport and mischief our forte, but doomed to academic failure. Granny found her vocation when embarking on nurse training at the Royal Victoria Infirmary - leading to becoming a District Sister, her favourite patients residing in the Byker Wall.

The Gateshead Millennium Bridge (The Blinking Eye), another Tyneside icon. Immediately left of the bridge is HMS Callliope (Royal Naval Reserve). This features in both our families' histories. Granny's father Donald (your great grandfather) joined the Navy once of an age to sign up, and served on Motor Torpedo Boats – taking part in D-Day.

Calliope played a role in my mother's family story. Your great granny (Paddy) served as a Wren Officer and married a young submariner who tragically lost his life before their first anniversary. Her elder brother served as a submarine commander in the Atlantic, Mediterranean, and Pacific - he survived the war and was highly decorated. Her brother-in-law captained a destroyer. Calliope provided an important venue for them, both during and after the war.

It would be remiss not to mention Mum's younger brother, elder sister and their father (my grandfather) who served in the army. The family made the front page of the Journal/Chronicle regarding their war effort – the Fighting Andersons. It's hard to compute what our relatives encountered and endured at such a young and impressionable age.

There is more to Newcastle's quayside than the river frontage – Bessie Surtees House dates back to the 16th and 17th century and a rare example of Jacobean domestic architecture. This part of Newcastle is fascinating and well worth exploring the many passageways, alleys and stairways – a valued aspect of the City for Granny and me.

It'll be hard for you to comprehend how different this trendy part of Newcastle was even in our late teens. The river an open sewer and stank – when crossing the High Level bridge in summer, it was advisable to close your car windows. The restoration of the river and return of salmon is nothing short of a miracle.

Further along the street from Bessie Surtees House are a couple of venues that Granny and I knew in our younger days – Julie's nightclub and the Cooperage. The Cooperage a particular favourite of ours – reminiscences of hosting a 'welly hoying' competition out in the front street. Many memorable nights and escapades in this part of town.

The Castle Keep should come as no surprise given the name of the city - but many are unaware of this. On the other side of the railway track is the Black Gate (the ancient entrance to the city) where friends of mine once lived and I frequented on many occasions in my youth – now a museum. Memories of Stefan playing his Northumbrian pipes.

A great view looking downriver, and it encapsulates just how the quayside has transformed from industry to leisure – the Sage and Baltic bear testament to this. There is so much more to Newcastle, but sadly space is limited.

The final picture comes from Jesmond Dene, an amazing oasis in the middle of the city. The Dene featured in Granny's formative years, and also when we lived in Gosforth. It's also a place where your great grandparents took our youngsters for adventures through the woods and across the stepping stones – as well as visits to Paddy Freeman's.

We could write more about this amazing city: my time as a student, with very little effort put into studying. Living in a house with an outside netty. Your Granny's high party life in the 70s. My first and ill-fated business venture. Our long courtship and story. Fortunately, or unfortunately, we've run out of space and so a tale to be told another time.

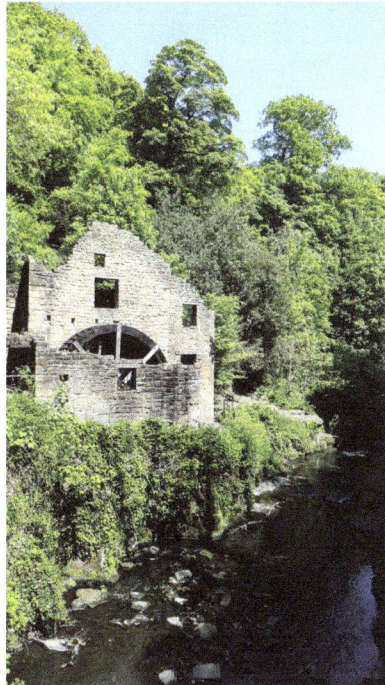

Authors & Books
Battle for the Atlantic – Jonathan Dimbleby

A book telling the horrors of war in the Atlantic, provides an insight as to the debt of gratitude we owe our ancestors (your great grandparents' generation) and the freedoms we're privileged to live our lives – make the most of it.

Musicians & Popular Music
Jimmy Nail and Mark Knopfler – Big River

Big River tells the story of the mighty Tyne. Worth checking out Jimmy's portrayal in the TV series 'Auf Wiedersehen Pet'. Brings back memories of my days in construction - a number of our lads having worked on sites in Germany.

Composers & Classical Music
Elgar – Nimrod from the Enigma Variations (Full Orchestra or Organ)

A fitting piece of music to remember our family's story. The organ composition takes me back to Remembrance Sundays with my parents as a youngster. Extremely apt as Donald (your great grandfather) died on the 11th day of the 11th month - last year. This week we spread his ashes and planted a tree in his memory.

Fine Wine
1995 Château Mouton Rothschild, Premier Cru Classe, Pauillac, Bordeaux, France

Granny's sister, your great aunt, lost her husband (your great uncle David) to a brain tumour 7 years ago. It's fitting we're drinking one of David's favourite wines from his cellar, to remember both him and his father-in-law (your great grandpa Donald), who was also partial to a good red wine.

TOWNSVILLE & ORPHEUS AUSTRALIA

Our last trip to Australia goes back to early 2018. We had planned to attend an important 70th Birthday celebration in Melbourne at the beginning of this year. But COVID-19 put the kibosh on that and there's no clear indication as and when that may be a possibility. So, it's cathartic to look back on a memorable trip and catching up with great mates.

We flew from Edinburgh to Dubai, and after a 4 hour stopover went on to Brisbane, an even shorter stopover and on a flight to Townsville. We'd no time for jetlag as our first night involved a stay at the Ville Resort-Casino and meeting the Aussies for dinner, ending up in the casino as Granny and Cousin Penny fancied a flutter.

We weren't in Townsville long enough to learn much about it, other than via our friendly taxi driver telling us about their recent devastating floods and showing just how high the water had been. Another interesting snippet of which I was to learn more at a later date: he pointed out a painting on the cliffs, referring to it as the Tall Man.

Our mode of transport the next day to Orpheus was a first for Granny, and of a much higher calibre than any I'd previously flown in. The trip out to Orpheus was spectacular, the weather perfect. Granny was completely hooked, on both the helicopter and our pilot. My bank balance sadly not up to this becoming a regular mode of transport.

As my picture demonstrates, Orpheus Island Lodge is paradise. The island is a National Park and the only other human inhabitants are situated at the far end, at a research station overlooking Hazard Bay, run by James Cook University. It's remote, beautiful and we were treated like Kings and Queens – a truly spectacular experience.

The food, wine, accommodation and service was up with the very best. Granny and I would like to take this opportunity to thank all those involved in running such a slick and friendly operation, and making our stay so very special. It does help when they're in a position to turn on night skies as per my picture – breath-taking.

During our few days at the Lodge we learned Orpheus Island is 12 kilometres long and part of the Palm Island group. The adjacent islands are Pelorus and Hinchinbrook – the larger island across the water Palm Island. Orpheus is roughly 80 Ks from Townsville and the nearest mainland township is Ingham 20 Ks away. Beautifully remote.

We had a number of memorable excursions: snorkelling on the Barrier Reef, canoeing through the mangroves, a castaway picnic on the uninhabited island of Pelorus, a low-tide educational beach walk. It would be hard to say all our time was taken up expanding our knowledge – swimming, relaxing, drinking and eating played their part.

Orpheus Lodge is famous for its culinary experience 'Dining with the Tide', which takes place at the end of the jetty, a maximum of 4 - great food and drink. Unfortunately when our time came a cyclone hit and we had to re-convene inside. The wine was exceptional, of which you will hear more when recording this week's entry.

I encountered this local resident on our castaway picnic on Pelorus. Granny being arachnophobic was not informed until viewing our holiday images. I assured her they were only native to Pelorus - so hopefully we can return to Orpheus one day. We had a most unusual lunch venue in an abandoned house overlooking the sea. There were plans to build an amazing new secluded lodge on the site – COVID and the pandemic may have scuppered that.

My education regards Australia has come late in life and remains sketchy. On the night of 'Dining with the Tide' I learned a bit about the Cathy Freeman Foundation and aspects of their work on Palm Island. I was aware of Cathy Freeman, the iconic Olympian runner who won the 400 metres at the Sydney games in 2000. For those about at that time, who can forget her lap of honour draped in the Aboriginal and Australian flags? An inspiring moment.

Sitting in the lap of luxury – the story of Palm Island and the Indigenous people certainly makes one stop and think. I was brought up in a household where respect for every race, creed and colour was taught – an important lesson. My father (your great grandfather) fought with the Gurkhas in WW2, from whom he gained much wisdom.

My final image shows Palm Island in the distance and the helicopter departing. Unfortunately we were not on board, as the cyclone ensured a fairly choppy boat ride was our only option - before taking a taxi on to Cairns.

Authors & Books
The Tall Man – Chloe Hooper

A recommendation from staff at a favourite bookshop in Albert Park (Melbourne). Not comfortable reading, but it provides an insight to Palm Island and the life and death of one of its indigenous inhabitants – Cameron Doomadgee.

Musicians & Popular Music
Dolly Parton and Kenny Rodgers – Islands In the Stream

The title and words relate perfectly to a romantic island, as well as to Granny and I – brings back memories of a memorable trip and tells a little of our relationship 40+ years on from walking down the aisle together.

Composers & Classical Music
Christoph Willibald Gluck – Melody from Orpheus and Eurydice (Evgeny Kissin on Piano)

Only one possibility for this week 'The Melody from Orpheus' – couldn't be more fitting.

Fine Wine
2015 Tolpuddle Vineyard Chardonnay Coal River Valley, Tasmania, Australia

A memorable magnum drunk with great mates on the night of 'Dining with the Tide'. Funny, the following evening the bottle was duly brought to the table and less than a glass remained – hence a 2nd magnum required.

WEEK 24
NORTH
NORTHUMBERLAND COAST
ENGLAND

June

19th was the day lockdown restrictions were to be lifted. Unfortunately the Indian/Delta Variant has generated a spike in COVID-19 cases, so Freedom Day has been set back a further month. The positive being the number of people being vaccinated has far exceeded expectations, and hospital admissions and deaths remain low.

Travel to faraway destinations remain on hold for Granny and **me,** so fitting to demonstrate what's on our doorstep.

I took this picture 20+ years ago which provides an indication of the amazing secluded beaches on Northumberland's beautiful Coast. The view remains the same today, the style since replaced and not as photogenic. Cocklawburn Beach is but a few short miles from Berwick-upon-Tweed, and can be accessed by a magical walk along the cliffs.

The more normal mode of travel is by car via the village of Scremerston. In my youth Cocklawburn was a great venue for picnics, later in my teens well-known for late night parties, and, reverting to old fashioned terminology, 'Courting'. Granny and me in our early days frequented Cocklawburn, and to this day regularly walk along this special beach.

I thought long and hard as to whether I should include the following: cars and Cocklawburn have form with me and a misspent youth. This long before the current eminently sensible attitude and laws relating to drink driving, driving fast and under the influence of alcohol a common occurrence at that time. The inevitable consequence: danger and accidents. I rolled my Mini at this location, fortunately without inflicting injury on others or myself.

Later I took up racing cars on the track, not the road – a better place to explore yours and a vehicle's limits.

Further down the coast is Holy Island (Lindisfarne), a magical and ethereal place when not overrun by mass humanity – as a proud Northumbrian, I do like my space. This picture demonstrates 2 pilgrims walking the original method of accessing the Island, the Pilgrim's Way. Lindisfarne is cut off from the mainland when the tide comes in.

The causeway provides vehicular access and was built when I was just a baby – prior to that, the journey via a couple of dedicated taxis following the Pilgrim's Way and over the sands. It amazes me the causeway still catches out many unwary drivers who fail to check the tide tables. An expensive experience, as it inevitably leads to the loss of a car.

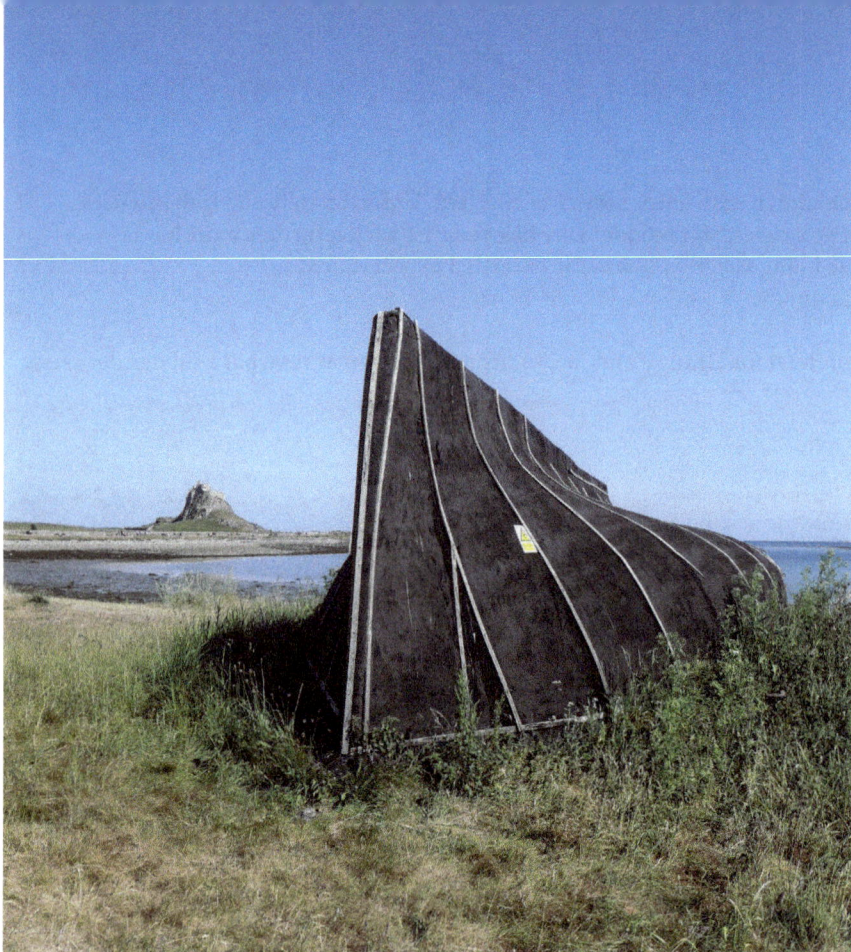

Holy Island was a regular haunt in my youth, as we frequented several of the Island's pubs – one in particular, the Crown and Anchor, run by a good friend. Many memorable nights of entertainment on Holy Island. Later in married life, Granny and I partook of excellent food and wine at the same venue – a hostelry now run by a friend from my school days. Our son (your dad or uncle) spent a couple of summers working there as a waiter – great experience for his future career.

Lindisfarne has a fascinating story to tell from early Christianity and the Vikings. Much more able scholars than I have written about St Cuthbert, the Priory and history surrounding this special place. Better you explore their scholarly works. My image of Lindisfarne Castle is a recognisable landmark all the way to Berwick. Once again, much has been written about this and its transformation to a private dwelling by reknowned architect Sir Edward Lutyens in 1901.

As ever space is limited so I'll leave you to explore the history of this unusual and special place. Before moving on, this picture looks south to Bamburgh, whose castle silhouette can be made out in the far distance.

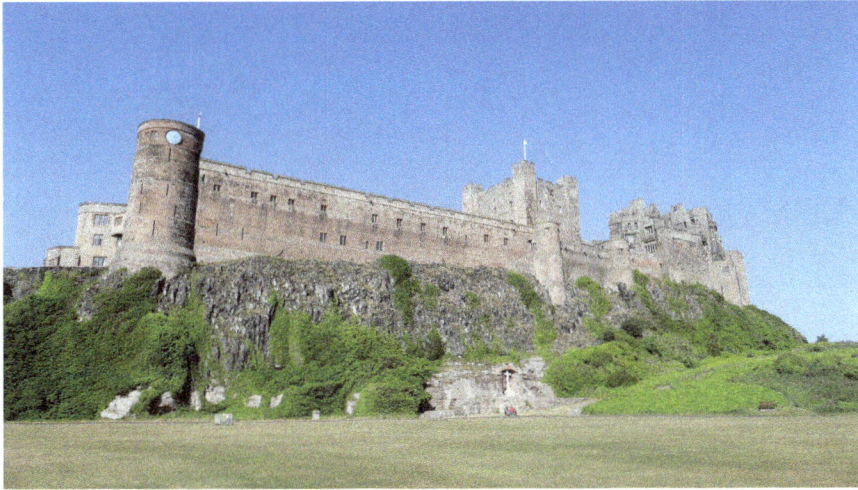

Bamburgh Castle dominates the village and as per Holy Island its story and history is well documented. A lesser-known history relates to my family's involvement with Bamburgh. My grandmother, uncle and aunt (your great, great grandmother and great, great uncle and aunt) are buried in the churchyard at St Aidan's.

Sadly my grandmother died before I was born, and I've little knowledge of her life. Your great, great uncle is someone whose history is worth examining – a highly decorated Black Watch officer in WW2, later went on to help set up a National Park for the Sultan of Brunei. I'll leave you to explore his extraordinary story.

Bamburgh has many happy memories as a child: the beach, egg pool, picnics, playing cricket in front of the Castle, the Bamburgh Tennis Tournament. Later attending special events, weddings and funerals – memorable moments.

Authors & Books
The Last Kingdom Series – Bernard Cornwell

The 13 books in the series tell the story of Uhtred of Bebbanburgh - being the Saxon name for Bamburgh. Provides a fascinating insight to the area and its history.

Musicians & Popular Music
Steppenwolf – Born to be Wild

A great favourite of mine at that time - featured in the film Easy Rider (1969). A period of my life when your great grandparents must have despaired of me, having recently broken various bones in a motorbike accident.

Composers & Classical Music
Andrew Lloyd Webber – Pie Jesu (Requiem)

A choral piece seemed right in relation to this week's locations. Andrew Lloyd Webber composed and dedicated this to the victims of terror in Northern Ireland. Chimes with aspects of Lindisfarne and Bamburgh's story.

Fine Wine
Krug, Grande Cuvée, 167eme Edition, Brut, Champagne, France

A special bottle of champagne given to me by our 2 amazing children to mark a momentous occasion: This week I stepped down as Chairman of the company my business partner and I founded 40 years ago.

Week 25

Granny failed to get her **second** helicopter ride: **it** should have been from Orpheus to Cairns – but due to a cyclone, a bumpy boat ride and taxi in its stead. The plan **was** to pick up a 4WD in Cairns and drive to Mount Mulligan.

Once in a blue moon or lifetime, you may be lucky enough to receive an invite to sample the delights of an amazing new venture in advance of its launch. Our trip to and from, as well as our stay, was very special and memorable. This image provides an inkling of our surroundings. Magical, and somewhere we'd love to return **to** one day.

I'm getting ahead of myself, in that the drive from Cairns to Mulligan was also special. Unfortunately, time and weather were against us - so no photographic evidence. But take it from Granny and me, it was fascinating. The steep and winding road from Cairns, through rainforest to the vast plain above, made our vehicle work for its keep. The contrast of the flat plain and impressive agricultural landscape are images I'd have liked to capture.

The landscape changed once more on entering the gates to Mount Mulligan. The cattle station covers 70,000 acres – the 2.5 hour drive on an un-surfaced track in the aftermath of a cyclone and torrential downpours made for an interesting journey. Memories of a major boulder in the centre of the road painted bright blue – to save one's sump.

I'm pleased to report this is not how our Mitsubishi 4WD ended up after the trip – but it highlights the problems a previous unwary traveller must have encountered on the road to Mulligan.

Prior to our arrival at Mulligan, Granny and I had no concept of what we were in for. Our Aussie mates vie with us in arranging Magical Mystery Tours. Their description: remote camping, and totally on our own-some.

On arrival we were blown away – this our accommodation. If this is Aussie camping - bring it on. The facilities in each individual unit (my image shows 2) are absolutely stunning. The size, style and facilities up there with the best, anywhere. Our cooking and dining facilities an enormous and beautiful contemporary structure supported by vast reclaimed timbers from Sydney harbour.

Camping never high on Granny's agenda – so we were absolutely blown away and in 7th heaven.

As you may have gathered, the 4WD vehicles outside our accommodation turned out to be for our use, and we made the most of these. They were amazing fun to drive and Granny became very proficient and addicted. This image demonstrates not only the vehicles - but the beauty and vastness of the remote landscape in which we found ourselves.

Waking up on our first morning to the view demonstrated by my first image and the sound of exotic bird song – magical. The adventure planned for that day was a visit to the township of Mulligan – now a ghost town. The only original building left standing and occupied – the hospital, which was being re-purposed for staff accommodation.

Mulligan was established in 1912 and abandoned 46 years later in 1958. The town was developed to support a coal mine. Mulligan has a tragic tale to tell: on the 19th September 1921, seventy five people lost their lives due to an explosion below ground - 25% of the town's population. Queensland's worst ever mining disaster.

A visit to the graveyard emphasised the lives lost on that fateful day. I wonder how long it took for the news to reach relatives in Oz and the other side of the World. Puts aspects of the current pandemic and lockdowns in perspective.

As ever, not enough space to record our many adventures in detail. But this image demonstrates a memorable trip to the waterfall descending from Mount Mulligan. An interesting interlude on the way related to a rock-fall in a railway cutting blocking our path. Utilising the winches on our trusty 4WDs enabled us to continue unimpeded.

Other escapades related to barramundi fishing, canoeing on the lake, the search for a lost goldmine, the infamous dunny can hunt, and others. A multitude of firsts for us - a truly unique trip, and an experience we'll never forget.

My final picture aptly demonstrates the working nature of Mulligan as a cattle station.

Authors & Books
On Leopard Rock – Wilbur Smith

For many decades my holiday reading involved the purchase and devouring of Wilbur Smith's latest book. On Leopard Rock paints a picture of his many and varied adventure stories. Mulligan would make a fitting location.

Musicians & Popular Music
Cat Stevens – Morning has Broken

A favourite for both Granny and I. Takes me way back to my time living off the New Kings Road (London) in the early 1970s. Perfect title choice in relation to waking up each morning and looking out over Mulligan's magical landscape.

Composers & Classical Music
JS Bach – Cello Suite No 1 in G Major

A composition engendering all the many different moods and emotions of Mount Mulligan's story and history, as well as our catalogue of exciting and wide-ranging adventures.

Fine Wine
2009 Domaine de Trevallon Rouge, Alpilles, Provence, France

No record of what we drank at Mulligan, even though we'd access to the owner's wine cellar. In its place, a bottle consumed on a memorable trip to Berry Bros. and Rudd in London and dinner at Café Morano – as a thank you.

WEEK 26
CHILLINGHAM & CRAGSIDE ENGLAND

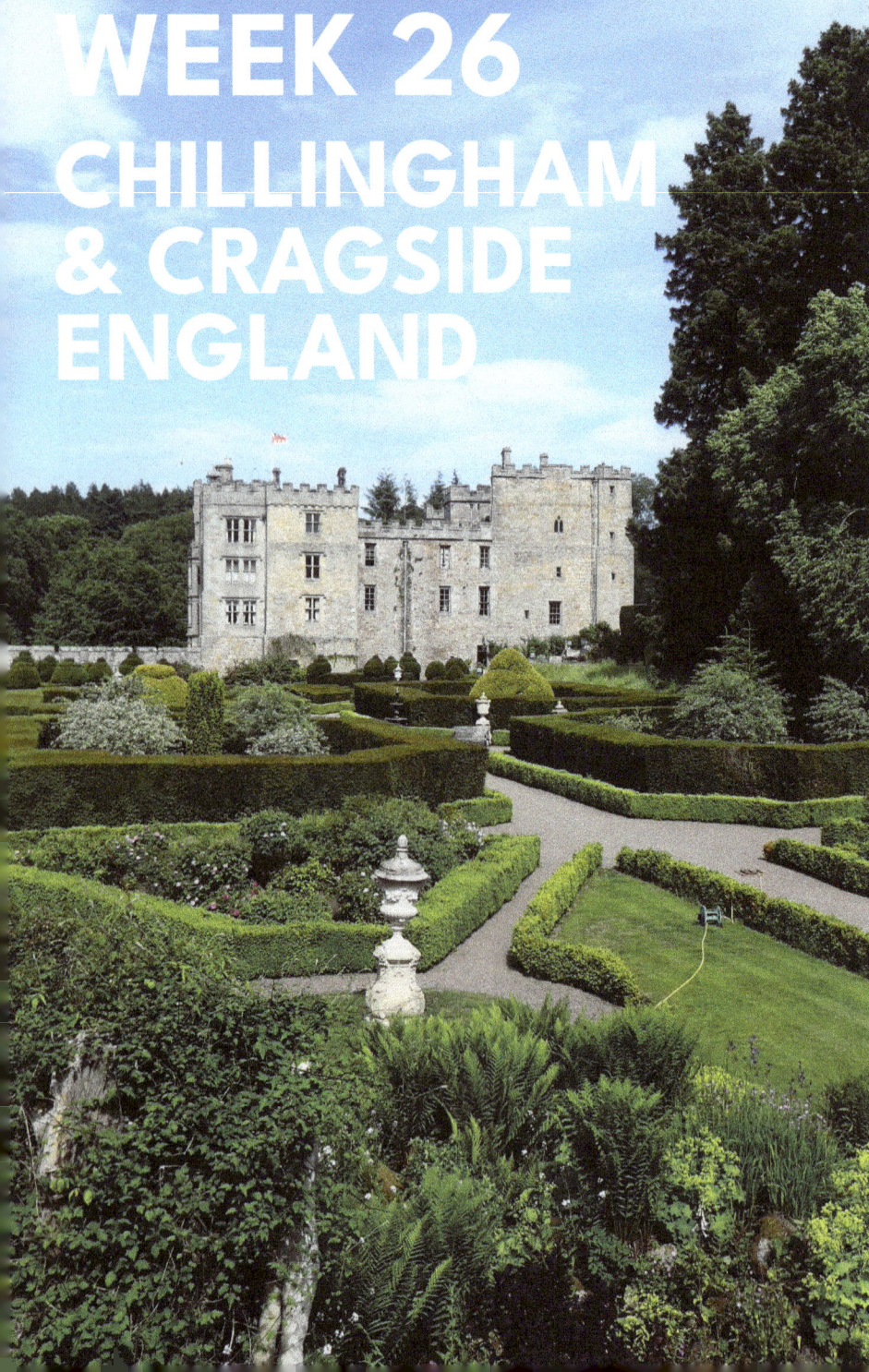

A major milestone has been reached with this week's Sabbatical entry - the halfway stage.
Possibly a good time to reiterate and re-evaluate Granny and my thoughts when embarking on our odyssey at the start of the year? The message to our Grandchildren was simple and succinct:

The World's Your Oyster

Do with It What You Can

With an Open
& Enquiring Mind

Eighteen months ago, the pandemic came out of nowhere and impacted all our lives. Not all negative - it's made Granny and me ruminate on how lucky we are to have our family and friends, a nice house and garden in a beautiful part of the country, managed to maintain our livelihoods, no home schooling, and our health – amongst others.

Chillingham and Cragside epitomise two completely different aspects of the Sabbatical. The first being the development of friendships and memories. The second relates not only to a spectacular place, but the people who made this possible. Both locations are within 40 miles of our home; COVID-19 has made us better appreciate all that's on our doorstep.

Granny and I recently made a pilgrimage to Chillingham (on our way to Cragside), as it aptly demonstrates the importance of friends, fun and memories. 30 or so years ago we attended the function of our lives. We partook in a gala charity ball held at the Castle. A spectacular and memorable event, and one still talked about to this day.

We started with champagne in our garden, and a bus on to the main event, the weather spectacular, and dropped off at the Castle steps. On entering the inner courtyard, served more champagne and regaled by a full military pipe band. Dinner and dancing took place in a marque on the lawn. Other entertainment involved a casino in the Castle, laser-quest in the dungeons, a full dodgem car ride in the woods, and an enormous bouncy castle.

We arrived in daylight, left in daylight, and onward to a breakfast, and later a lunch. Oh to be young again.

Chillingham Castle is well worth a visit if only to marvel at the tenacity of its owner Sir Humphrey Wakefield and family who've taken on the Herculean task of attempting to restore a rundown ancestral home. Makes Granny and I very happy to live in a much more compact and manageable house with all our mod-cons.

Cragside is much younger, and purchased as a wild and desolate moorland overlooking Rothbury in 1863 by Lord and Lady Armstrong. The scale and ambition of what they created not only at Cragside, but in life, is totally mind-boggling. This man was a colossus not only within Britain, but the World – someone deserving of much wider recognition.

He started life as a lawyer, was a self-taught scientist and inventor, an industrial magnate on the world stage, an educationalist, a philanthropist, an environmentalist and conservationist many years before those words reached the importance and prominence of today. To all of this you have to add the development and creation of Cragside.

It's interesting, the coincidences and relationships to other Sabbatical entries. We've recently featured Bamburgh Castle (Northumbrian Coast) and Jesmond Dene (Newcastle-upon-Tyne). The restoration of Bamburgh Castle (at vast expense) was carried out by Lord Armstrong after the death of his wife of 58 years. The pair also built Jesmond Dene House and created an amazing inner city landscape, later donating the Dene to the citizens of Newcastle.

Their philanthropy has few equals and encompasses Newcastle University, the Hancock Museum, the Royal Victoria Hospital to name just a few. I leave you to look into their amazing lives and achievements.

To truly comprehend what the Armstrongs achieved at Cragside requires a visit. The house and grounds are spectacular. The first house ever to have electricity and light driven by hydro-power. Lakes, dams and waterways were created to make this all possible – how far ahead of the game were the Armstrongs. They planted over 7 million trees and shrubs. The original 6 mile carriage drive highlights the scope and amazing ambition for their project.

The final picture shows the road leading to the stable block and all the ancillary buildings. Granny and I early in our married life looked at developing the old laundry as a home – for various reasons, we never went ahead with this.

Authors & Books
Mao's Last Dancer – Li Cunxin

From Chinese peasant boy to prima ballet dancer on the World stage. An inspiring true story and a book I gave to a young family friend when facing up to major personal choices. She has more than risen to the challenge – brilliant.

Musicians & Popular Music
Bill Haley and the Comets – Rock Around the Clock

An extremely apt music choice and title given the magical memories of boogeying till dawn at Chillingham.

Composers & Classical Music
Isaac Albeniz – Asturias (Guitar)

The tempo rises and falls as well as appearing to come to a finale, only to find fresh impetus and continue on its way. Fits the bill regards the exuberance and staying power of youth and a very special night.

Fine Wine
2008 Veuve Clicquot Ponsardin Vintage Brut, Champagne, France

Chillingham was long before Granny developed her taste for fine wine – so a vintage champagne consumed on another occasion with our son and daughter-in-law at a memorable trip to Tobago in 2017. Possibly a good opportunity to remind my friends I've still not been reimbursed for the wine at Chillingham!

WEEK 27
HUA HIN
THAILAND

Granny

and **my first** ever Aussie trip involved a one-way airline ticket around the World and entailed no back-tracking. To which end we started in New York, flew on to LA, and then Melbourne. As you will have gathered, our Oz odysseys tend to be full on, and Hua Hin was the perfect rest and recuperation before returning home after a manic month.

Our final stop in Oz involved a few days in Sydney – Hua Hin proving the perfect counterbalance.

This is one of my favourite ever images. Granny and I had this picture postcard beach almost exclusively to ourselves, when I observed these 2 striking local ladies gathering clams – I also spied the kite-surfer. By some fluke, I managed to combine all 3 into the 1 picture. For me, it epitomises east meets west – extremes of poverty and wealth.

The previous evening we'd been collected from Bangkok Airport by a luxury limo supplied courtesy of the hotel. We were made aware of the discrepancy of poverty and wealth on our journey. The number of local Thai families on the move in the middle of the night, hanging onto totally overloaded trucks and motorbikes – made a lasting impression.

Travel is a chance to learn and realise many others don't have the same opportunities open to them as you do.

The heat and humidity at that time of year was intense – hence having the place almost to ourselves. A lasting impression for Granny and me: the lovely Thai people and their beautiful, polite, friendly manner and approach to life. Their greeting of Namaste demonstrates this, and apt in relation to the COVID-19 pandemic.

One of Granny's favourite places was to perch herself just above the beach, where there was a breeze - and waited on hand and foot. I've always had the total inability to sit still for any length of time. This picture demonstrates one of the swimming pools, which was a long winding continuous loop – I swam endless laps in the shade of the trees.

Granny and I, after an intense month, did not partake of leaving the luxury and relaxation of the hotel and its amazing facilities. But we did learn a little of Hua Hin's story. Adoption by the Thai Royal Family and the building of a Summer Palace, Klia Kang Won (place of no worries), in 1926, led to Hua Hin becoming the Thai Riviera.

We also learnt about the hill at the end of the beach – atop of which is a Buddhist temple, Wat Khao Takiap.

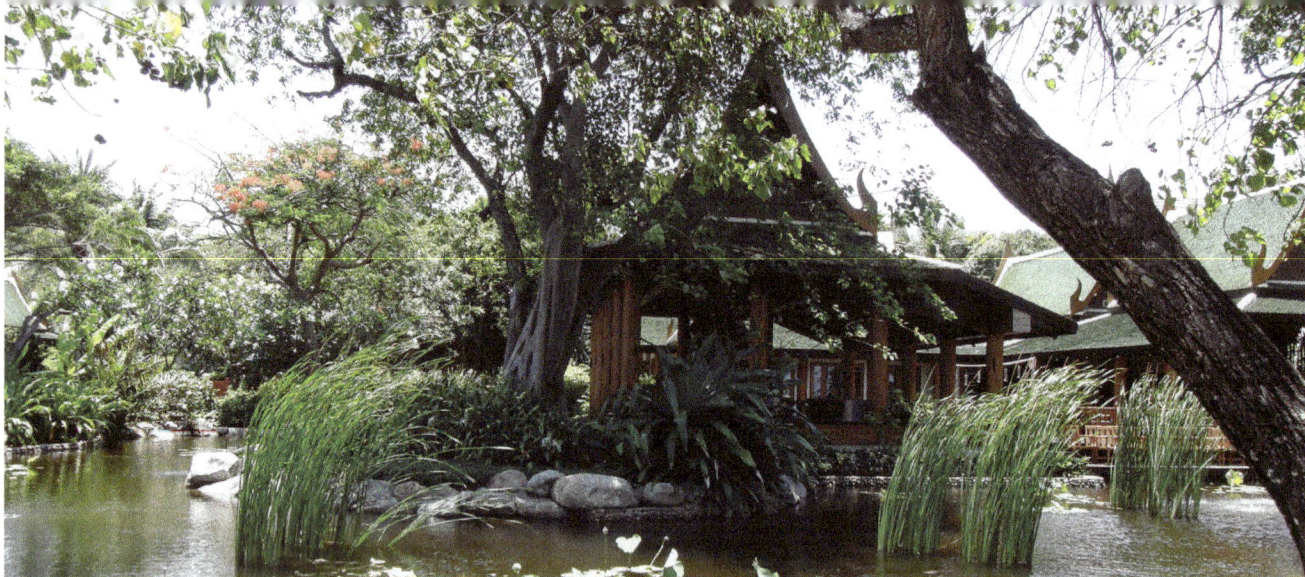

We had 3 full days at Hua Hin, so sampled all 3 hotel restaurants. This being the piece de la resistance; other than 1 other family, we had this amazing building to ourselves, and the staff treated us like royalty. We went for the fish extravaganza – everything fresh from the enormous fish tanks. Little did we realise this appeared to entail every single species of fish and crustacean available. A veritable feast, and one never to be forgotten.

A further memorable experience while at Hua Hin was a visit to the brand new spa, which had recently been constructed alongside the hotel. We partook of this amazing facility. The building, layout and architecture were spectacular. Best described as a journey along a never-ending corridor.

Granny and I had a treatment together – we floated off into 7th heaven, the most relaxing massage ever. The culmination being massive doors rolling back to reveal our own secluded sunlit courtyard, with a sunken pool. We were so taken with the whole experience, we booked in prior to departure on our final day.

This image demonstrates just how crowded Hua Hin beach was!

If you look very carefully you'll notice to the right, hidden in the trees, is a further restaurant and bar – McFarland House, a restored 19th century pavilion.

This is where our stay in the lap of luxury culminated. After our 2nd visit to the spa we had cocktails and a meal looking out over the ocean prior to departure.

A fitting end to a magical few days and memorable Aussie odyssey.

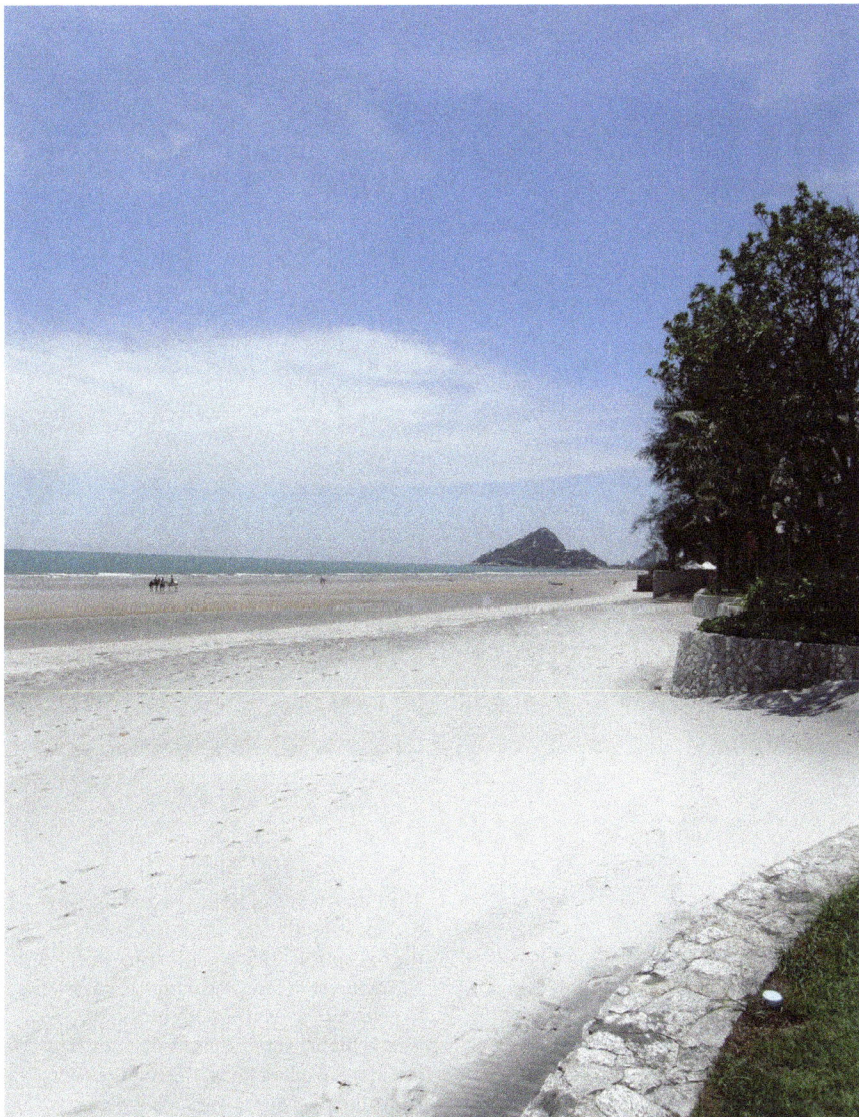

Authors & Books
Burma 44 – James Holland

COVID-19 brought about an unlikely super star, 100 year old Sir Tom Moore, who raised in excess of £30 million for the NHS. While Sir Tom was striding out, Myanmar (Burma) once again became a military dictatorship. Burma 44 provides an insight to the 2nd World War in this region, in which Sir Tom and many others played their part.

Musicians & Popular Music
Lou Reid – Perfect Day

Perfect title to return vicariously to the most perfect few days with my Bride of 40+ years (Granny) – extra special. Hopefully when this world pandemic is brought under control we may be able to pay a return visit.

Composers & Classical Music
Philip Glass – Violin Concerto – 2nd Movement

We gained a new grandson (Sandy) during the first Lockdown, who has just celebrated his 1st birthday. This piece of music we associate with him. Invokes total joy and rapture as per our memorable few days in Hua Hin.

Fine Wine
2017 Domaine Bitouzet-Prieur 'Les Corbins', Meursault, Burgundy, France

Beer and cocktails were the order of the day in Hua Hin – so a special wine brought out of our Cellar Plan to raise a glass and toast: to a time when travel and developing future memories are once more available to Granny and I.

WEEK 28
LOCHINVER
& ASSYNT
SCOTLAND

This
week marks the 1st time we've been able to holiday with mates since January 2020. All attempts over the last 18 months to so do ending in failure, due to Lockdowns and Government decrees. Official restrictions are being lifted this week, but infection rates sadly on the rise, hopefully the vaccination programme prevails.

The importance of holidays and making memories with friends a cornerstone of the Sabbatical. This week's trip came via an invite from longstanding special friends, to celebrate a belated 65th birthday. This is highly significant in that our two families have holidayed together over many, many years – our kids growing up as one.

There've been a multitude of memorable trips over a number of decades: Lyn Phail, Glen Tanar, Centre Parks, Glen Lyon, Portugal, Disneyland (Florida) – to highlight just a few. We've also had numerous adult only fishing, skiing, and sailing trips: Laxford, Lyon, and Tweed, Samoen, and Corvara, Croatia, Turkey, and Spain. A plethora of great memories.

This week a new adventure for Granny and I, as we're to fish 3 new rivers: the Inver, Kirkaig, and Garvie. Via a strange coincidence I'd previously spent a night at Lochside House, our accommodation. I was invited to a splendid and raucous dinner (without Granny) a number of years ago. The name highlights the magnificent setting.

It would be remiss not to mention the other couple in our party who introduced us to skiing, as well as accompanying us on various momentous fishing trips. Granny and I were extremely excited at being let loose once more.

We've never previously stayed in Assynt, but have driven past its magnificent landscape on our way to the Laxford many times. An infamous one being Granny's decision that we should take a short detour to see Achiltibuie; 3 hours later, we managed via a circuitous and torturous route to get back on track.

Day 1 (Sunday) involved an expedition up the Kirkaig to Fionn Loch for a picnic and a spot of trout fishing. The weather was overcast, but ideal for walking and fishing. Granny's watch recording a hike of 26,000 steps. Which impressed us, but overshadowed by our 4 younger companions following this with a 36,000 step trek the next day.

Fionn Loch is a beautiful setting, but the weather unkind regards photography. So a picture borrowed from our hostess, showing Suilven – we were unable to observe this due to being shrouded in cloud. Our picnic took place in Arthur's Bay. Named after our fellow fisher's mate - it relates to his ashes being scattered over the Loch via a rocket!

Our second day was spent fishing the Lower Kirkaig. A pretty piece of river if the water level wasn't at an all-time low. But Granny and I nevertheless had an exciting time fishing the sea pool at high tide; fish were leaping and splashing all around us. I briefly had one on – but unfortunately not to be.

Day 3 involved fishing the Garvie – lack of water meant our rods were never assembled. The sandy beach at the river mouth incorporates an interesting and unusual rock formation. A run up to the Laxford was more productive in that we made contact with various old friends from the many years we were lucky enough to fish this great river.

We'd been looking forward to Day 4, which involved fishing the Inver. Once again, the fishing Gods were not on our side due to a scarcity of water. We tried, but sadly to no avail - would love to return in better conditions.

A total change of tack on Day 5 – a trip to Handa Island, a nature reserve. A spectacular day and place. The 6 Km circuit around the island encompasses impressive vistas from cliffs and beaches, as well as panoramic views of the mainland and mountains. The birdlife was a spectacle to behold – we hadd a very special day.

Friday: our final full day, and wall-to-wall sunshine - 26 degrees. Three intrepid members of our party thrashed the Kirkaig's sea-pool at 5.00 in the morning – much colder at that time of day, but no joy. Granny and I took a trip around Assynt's Drumbeg road – stunning scenery when not concentrating on the undulating single track road.

The final morning provided brief excitement for our host at 6.00 in the morning. Fishing a hitch fly on the Inver, multiple salmon taking interest and 2 lost. A great finale to a magical and special week – we're on for a return visit.

Authors & Books
Gurkha – Kailash Limbu

American and British forces have announced their withdrawal from Afghanistan after 20 years. This book provides an insight to aspects of this brutal, unforgiving and unwinnable war. How will history recount their achievements or failure? Also a nod to my Dad (your great grandfather), a Gurkha officer in WW2.

Musicians & Popular Music
Boz Scaggs – Sierra

A tribute to a very special young lady who loved her many family trips to Scotland, and was very much in our thoughts during a magical week - at a worthy new location, now added to the Sabbatical list.

Composers & Classical Music
Arvo Part – My Hearts in the Highlands (David James)

A haunting piece of music which mirrors my and Granny's love for this amazing part of the World.

Fine Wine
2017 Storm Wines, "Vrede" Chardonnay, Hemel-en-Arde, South Africa

A wine shared with our host and hostess during difficult times – she professes this to be one of her favourites. Hence my last 2 bottles withdrawn from our cellar, to salute a memorable week with very special friends.

Granny

has put her foot down to make the point I'm not totally averse to visiting Cities – although my preference relates more to wide open spaces. Week 27's entry (Hua Hin) pointed out we'd been in Sydney the previous week. Our 1 and only visit to Sydney, and final leg of our original Aussie Odyssey. We loved it and covered a lot of ground.

Our hotel down by the waterside, my picture demonstrates our location, plus 2 amazing icons: The Sydney Harbour Bridge and the Opera House. The strangest of coincidences. We'd only just booked into our hotel, I was hot and bothered, and in need of an Oz tonic (a tinny or 2). So we took a walk down to the harbour area.

No sooner had we ordered a beer than I said to Granny: 'That's Sarah from Bamburgh'. Granny's reply: 'You're delusional; too much sun'. I paid no attention and chased after her. Turned out I was correct. Not only was it Sarah, but she was with another friend, Moira, from our next-door village in the UK. So a few extra beers consumed.

The Opera House fascinated me, as did how its perspective changed depending on whether viewed in daylight or at night. I attempted to photograph it from various locations, angles and times of day. This my favourite as it highlights the concept of a fully rigged sailing ship. We were unable to get tickets to the concert taking place, but happened to be walking past during the interval, so sneaked in for an impromptu tour of this amazing building.

We stumbled across an interesting museum relating to Aboriginal heritage, which by chance was close to our hotel. An enlightening exhibition for 2 ignorant Poms with very little insight to the history and suffering of the indigenous people of Australia. Not something we were taught in school – but something that should be on the curriculum.

We were slightly underwhelmed on paying a visit to Sydney's Botanical Gardens. Having recently been in Melbourne - Sydney's were not as impressive. The Sydney garden which stood out for us was the Chinese Friendship Garden, in Darling Harbour. A spectacular 3 acre extravaganza highlighting Chinese involvement in Australian history.

An image aptly demonstrating the best mode of travel in Sydney – water taxis and ferries. We took full advantage and visited numerous locations. The area and volume of water in Sydney harbour is on an immense scale; a truly amazing natural port, and an ideal way to explore much of this fascinating city.

This the mast of a replica of Captain James Cook's ship the Endeavour, on display at the Australian National Maritime Museum. Cook and his exploits are well documented. An interesting man who came from humble stock, his father a farm labourer from Ednam (Scottish Borders). The story of building the replica ship and its various voyages around the world is a further fascinating one, well worth delving into. We found our tour of the ship enlightening.

The National Maritime Museum also contains HMAS Onslow, an Oberon Class submarine. My uncle (your great, great uncle) was a submarine commander during WW2, Australia their base during the latter part of the conflict. Visiting the submarine gave a real sense of what he and his crew must have endured. An interesting fact relating to my uncle: I'm led to believe he commanded a captured German U Boat – possibly the only British officer to do so?

The Sea Life Sydney Aquarium provided Granny and me with a close-up of many of the fascinating sea creatures found around Australia's vast coastline. I'd snorkelled on the Barrier Reef off Cairns a few weeks earlier. Delighted to report I didn't come across one of these; thankfully this was my closest encounter.

A picture taken from the Heads, the entrance to Sydney Harbour, demonstrates this natural phenomena. We'd first travelled out to visit Sydney's famous Bondi Beach. The weather was not conducive to swimming, so we moved on to this amazing view. It provides a spectacular perspective to entering or exiting the harbour.

One further tale to tell reference our final night in Sydney. Our Aussie mates recommended a particular restaurant for the last evening of our epic trip, this perched at the very top of one of Sydney's tallest buildings. The restaurant, food, wine, and service were spectacular. But etched in my brain is a trip to relieve myself of a beer or 2. Whilst so doing, the panoramic view beyond the urinal encompassed the most spectacular view of Sydney's skyline.

Authors & Books
Wild Swans – Jung Chang

Our trip to the Chinese Friendship Garden makes this an apt choice. I read this book a number of years ago – it provides a fascinating insight into 20th Century China. A true story encompassing 3 generations of Chinese women.

Musicians & Popular Music
Katie Meluia – 9 Million Bicycles

Incorporates a nod to our Chinese theme, and an artist Granny and I much enjoy. We also subscribe to the sentiment.

Composers & Classical Music
Jules Massenet – Thais – Meditation (Nicola Benedetti)

A thought-provoking piece of music encapsulating the many and varied aspects in relation to our visit to Sydney.

Fine Wine
2015 Cape Mentelle, Cabernet Sauvignon, Margaret River, Australia

No concept of the wine drunk on our last night in Sydney. But a reminder of a far too brief visit to the Margaret River, and a very special wine tasting at Cape Mentelle - on the final day of our last Aussie Odyssey in 2019.

Granny

and I heard about Glen Lyon long before we ever set foot in this special place. My parents (your great grandparents) and friends holidayed at the Gate House to Meggernie Castle for many years. Members of the family joined them, but for whatever reason (work, too busy, lack of funds, whatever) I never managed to make it.

This was rectified 25 years ago, 6 months after my mum (your great granny) sadly died, and my dad (your great grandpa) was bereft having lost the love of his life. I rented a lovely house (Roro) way up the Glen near Bridge of Balgie. He and I had a memorable week: Fishing, walking, reminiscing, reading and cogitating - plus a drink or 2.

It did the trick – he then went on a holiday spree: a safari in Africa, a return to Paxos (Greece), numerous fishing trips and many others. As he said at the time, 'I needed to give myself a good talking to'.

For a number of years, Granny and I returned to Roro with friends and family and had many memorable holidays. One that stands out and proved a catalyst to our long and continued association with Glen Lyon relates to inviting our great mates, who originally introduced us to the Highlands via holidays in Glen Affric.

That week the weather was baking hot, not conducive to fishing. But there was major excitement in Glen Lyon as the twitchers were out in force, due to the sighting of a unique species of 'Great Tit' never previously seen in this part of the world. This memorable week led to our friends buying a property in the Glen – hence our ongoing association.

All these years and here we are once again in Glen Lyon with all our immediate family – only the second time we've all been together since the COVID-19 pandemic came about. Granny and I on cloud nine to have our family with us. This image from many years ago, highlights bluebells – something we've long associated with this magical place.

It's hard to distil quarter of a century of adventures, escapades and shenanigans involving Glen Lyon and Loch Tay. This image looking down on Kenmore at the eastern end of Loch Tay brings back a plethora of memories.

Granny and I have attended two amazing weddings in Kenmore's beautiful church, and the receptions back in Glen Lyon were spectacular and memorable. A further occasion etched on our brains relates to a particular night in a restaurant overlooking the Loch – better described and explained via this week's popular music choice.

A magical aspect of our stays relates to fishing the Lyon. This picture demonstrates my favourite spot, just below the falls and way down in the bottom of the gorge - the entrance to Glen Lyon. Granny and I love our fishing, and have had our fair share of success. Granny, as ever, wins the contest with an enormous 30lb+ salmon. To prove I've managed the odd fish, here's a picture of Burt returning a salmon I caught in my favourite pool.

When relating stories of Glen Lyon, it's important we mention the model village of Fortingall with its ancient Yew tree, Church, Glenlyon House, and Hotel. The parties and functions we've attended in the village over many years are too many to mention. But there've been some crackers, and the friendships Granny and I have forged deserve mention.

One particular function warrants highlighting: I invited 24 of my great mates to a dinner at the Fortingall Hotel to celebrate receiving my state pension. I thought it only fair to share this with friends, as mine came at 65, and they were having to wait till 66, 67, 68 or more. I prepared a brief 5 minute speech, which took 20 due to the heckling and barracking I received – extremely rude considering my pension was paying for their food and drink!

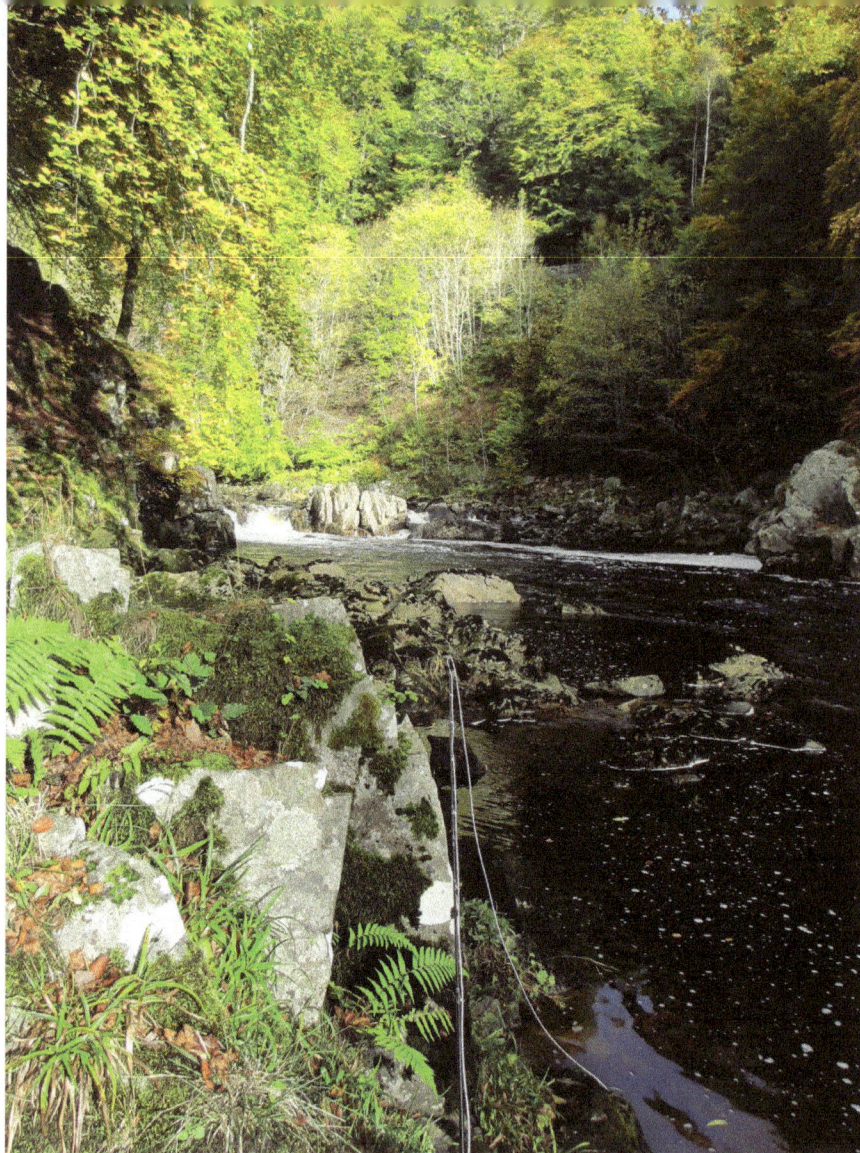

This picture highlights another of our favourite spots hidden way up in the hills where we picnic and trout fish.

Glen Lyon is very special to Granny and I, not only for its beauty and surroundings, but also the people. It would be remiss not to highlight the importance we place on the friendships we've been lucky enough to develop with such a wide and diverse range of people over the last quarter of a century. Even those who thought Granny must be my 2nd wife as she wasn't old enough to have offspring the age of our 2 wonderful children.

Thank you all – we've loved every minute and look forward (hopefully) to returning many more times in the future.

Authors & Books
Desert Flower – Waris Dirie

In-depth conversations on a multitude of diverse topics and subjects have taken place, with the Lady of the House in Glen Lyon - over many years.
This book will chime with our hostess. A demonstration of female achievement through unspeakable adversity and a myriad of topics that deserve to be aired. A pilgrimage worth emulating is coffee, cake and a book purchase at the 'Watermill' (Aberfeldy).

Musicians & Popular Music
The Proclaimers – I'm Gonna Be (500 Miles)

Relates to celebrating the wedding anniversary of special friends at a restaurant overlooking Loch Tay. We envisaged a quiet, sophisticated and secluded dinner for 4. Turned out to be the night of the monthly sing-along coach trip jamboree. After a few bottles of champagne we loosened up and led the audience in dancing to the above.

Composers & Classical Music
Rimsky-Korsakov – Scheherazade (Leif Segerstram)

Starts in a mild manner and involves a never ending series of crescendos – reminiscent of the many and varied escapades we've been involved with in Glen Lyon over the last quarter of a century. Too many to recount.

Fine Wine
J.M. Seleque, Solessence, Champagne, France

A wine supplied by our son, the fine wine specialist - vintage and memorable on so many counts. Having the whole family in one place for only the 2nd time in almost 2 years – very, very special for Granny and me.

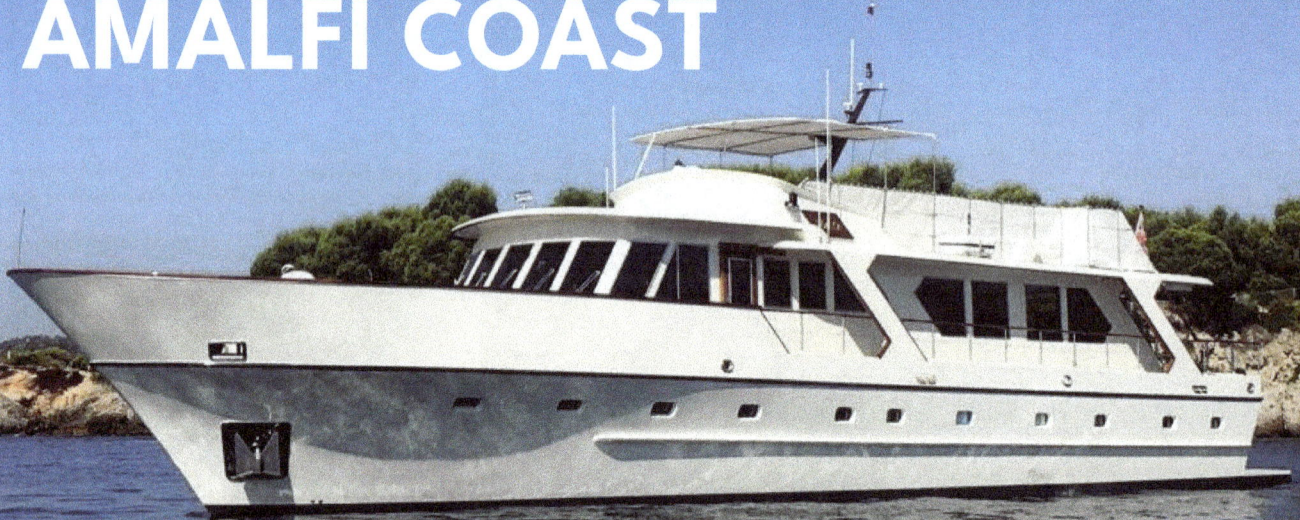

The Sabbatical paints a picture, Granny and I have led a charmed life - exotic destinations having been the norm. This week's entry could well add to this, but misrepresents our story. These 3 adventures cover a period of 60+ years.

It's important to understand Granny and I squandered our education opportunities - in both cases coming to a premature end. Granny made up for this by excelling in her nursing career. I on the other hand went to London at 19 to make my way in the World. At 24 years old I lost everything in an ill-judged business venture. Granny (bless her) married me 2 years later – our story relates to much hard work, and exotic holidays a long time coming.

The 1st adventure goes counter to this tale. My uncle (your great, great uncle the WW2 submarine commander) led an interesting life – at one stage in the mid to late 50s/early 60s he made a living captaining a luxury motor yacht, the 'Swiss Miss'. My Mum and Dad offered a last minute subsidised deal due to a charter cancellation – I was 7 years old.

My parents rallied a group of family and friends and off we went. We travelled by car to London, part of this on Britain's first motorway (the M1), which had just opened. I've a vague recollection of an early morning picnic in St James' Park and seeing Buckingham Palace. We flew from the recently-opened London Airport – later renamed Heathrow. Our flight in an unpressurised DC3 Dakota – a 2nd WW plane. A great adventure for us youngsters.

I should add it was more than a decade later, before crossing the channel once more – on a hovercraft (ferry) – and almost a further decade before another flight in an aeroplane. Exotic holidays not the norm for our family.

Sadly no imagery of the 'Swiss Miss' and our epic trip. We sailed from Nice and visited St Tropez and Portofino – I'm reliably informed we berthed alongside Brigitte Bardot's yacht (the pin-up film star of that era). We sailed to Corsica and I've recollections of seeing the USS Forrestal, an enormous American aircraft carrier. Other than eating raw sea urchins off the harbour wall with Pedro, a Spanish civil war exile (the boat's engineer), my memories are scant.

The point I'm attempting to make is: exotic holidays have not come about due to being born with a silver spoon. A much more potent weapon relates to Granny being the perfect house guest. I made mention of this in a speech to 24 great mates at my 65th birthday party, which took place in Glen Lyon a few years back. Here's a short extract:

I'd like to take this opportunity to acknowledge your patience and help in assisting my wife in developing my skills as a 'Professional House Guest' – this is much appreciated and has played a vital role in my ongoing quest to become the model citizen she so desires. I should add my apprenticeship still has a long way to go.

Which brings me to 2 further memorable trips on luxury motor yachts. The first of these took place at New Year some 40+ years after the 'Swiss Miss'. Great friends invited us to join them in Puerto Pollensa (Majorca) aboard the 'Annie Girl'. An unforgettable holiday and the earlier picture of the night sky demonstrates a spectacular evening. This next image of their mad mates buzzing the boat provides an insight to the quiet time encountered on our trip.

Before moving to our third voyage I'd like to highlight an anchorage which brought back memories. Many years previously on a holiday with our 2 kids, we stayed at the infamous 'Stalag Mosqueda' – a horror show. I remember a line from my letter of complaint: - 'I don't expect to have to listen to my neighbours urinate, defecate and fornicate'.

The picture at the start of this week's Sabbatical entry shows 'Stalca' – a magnificent vessel restored by great mates. She was originally built in 1971 for Princess Grace of Monaco and named after her 3 children – Stephanie, Albert and Caroline. We've massive admiration for all the work, effort and dedication our friends put into renovating 'Stalca'.

Our trip on 'Stalca' 4 years ago, along the Amalfi coast, was stupendous – how the other half live. Three crew looking after

6 guests – Granny and I could grow to living the life; unfortunately our bank balance is not up to this. These final pictures show just a couple of the amazing places we dropped anchor on our voyage – very special.

The weather went against us at the end of our trip, which created the opportunity to visit Pompeii and Herculaneum. If ever you get the chance, go and wonder at a society so far advanced and sophisticated for their time.

Authors & Books
Like A Virgin – Richard Branson

As an uneducated philistine I'd little option other than to make my own way in the World. Richard Branson's story provides an insight as to what is possible. My interest in money, power and influence falling way short of his.

Musicians & Popular Music
The Doors – Riders On the Storm

This piece of music takes Granny and **me** back to our teenage years. Fits this week's Sabbatical entry in relation to a storm sending us back to port, and hence our introduction to Pompeii and Herculaneum.

Composers & Classical Music
**Edvard Grieg –
Piano Concerto in A Minor**

Only one choice for this week's entry: on a previous visit to Naples my brother arranged a visit to the Opera House – a stunning and amazing building with a fascinating history. Grieg's piano concerto was performed and mesmerised Granny and **me**. Totally apt in relation to memories of the French Riviera, Majorca and Amalfi Coast.

Fine Wine
**(Vintage unknown) Casa d'Ambra
La Vigna dei Mille Anni Ischia, Rosso,
Campania, Italy**

The Sabbatical relates to memories, and here a wine recommended on a short break to Rome, many years ago, when our waiter (an Ischian) took a shine to one of our party and introduced us to this. At the time Granny and I'd no idea of where Ischia was located – a magical place, as demonstrated by the image highlighting their fortifications.

WEEK 32
SOUTH DOWNS & EAST SUSSEX ENGLAND

Granny

and I first became aware of Berwick (East Sussex) when our son began dating a certain young lady who hailed from that neck of the woods. This led to a conversation while fishing on the Laxford reference our son wanting to ask said young lady to marry him, and not having the necessary funds to buy a suitable engagement ring.

This was resolved by providing a ring left to Granny by my Mum (your great granny). The proposal for her hand in marriage took place in the Lake District on a family gathering. The explanation as to the ring led us to understand our son had made a great choice. His bride to be stating: "I would have been equally delighted with a Gummy Ring".

Granny and I's first visit to Berwick (East Sussex) took place 2 days prior to their wedding. Slightly fraught as much was yet to be done, our son flying in from New York where he was working at that time - the day before his wedding! Getting a puncture on the return journey from Gatwick, amongst other things, didn't exactly help the situation.

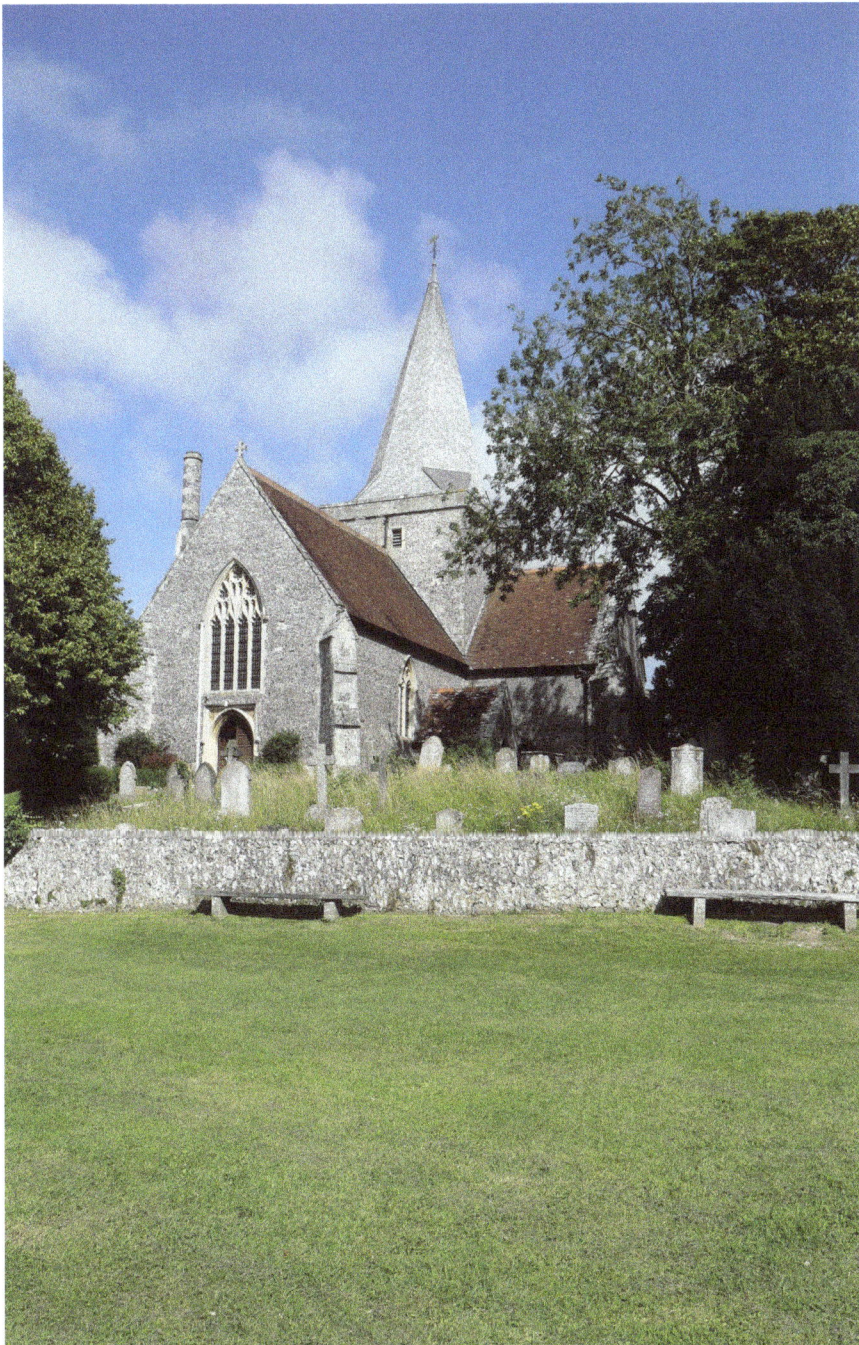

If we hadn't already utilised 'A Perfect Day' by Lou Reid in a prior Sabbatical entry, it would be ideal for this week's popular music choice. The service was held in the pretty little church at Selmeston, a small village adjacent to Berwick. The numbers restricted due to the church accommodating a maximum of 100 – so a full house.

The service was memorable and joyful. Best summed up by my conversation with the organist and vicar, who explained this being only the second wedding they'd held in the church in 2 years. The organist's comment he's normally asking himself: 'Hello is anyone out there, when hymns are sung' - not in this case.

The participation of those in attendance and the sheer joy encapsulated within the service - never to be forgotten. A certain participant took his involvement extremely seriously and brought the house down when the vicar asked a rhetorical question: 'Is anyone listening out there?' Our young Grandson (JJ) leapt to his feet and stated 'I am!' His other star turn was to drop the ring which he'd been entrusted – the ting-ting on the stone floor heard by all.

The reception in the beautifully-decorated village hall an equally joyful and memorable event. The speeches, music and dancing - as well as the copious quantities of excellent food and wine made for a very special day.

The South Downs and East Sussex have been incorporated into the Sabbatical due to subsequent but more recent trips to the area, and the belief they'll play a more substantial role in the future. The reason being our son and daughter-in-law moved to the Parrish of Wilmington (not far from Berwick) 2 weeks after our government introduced the first Covid-19 lockdown. Their previous home being in Peckham (South London) – excellent timing.

Another important point to make: we now have a new Grandson, born in June of last year – a pandemic baby. Lockdowns and government decrees meant we've had little opportunity to visit our son, daughter-in-law and grandson in their new home – thrice, to be precise. But we can already see and appreciate the area's attractions.

We managed just one visit last year, as every time we tried to arrange something, new restrictions were imposed. The most recent trip (last week) being our third. We were called on to help out with babysitting due to our son starting a new job and our daughter-in-law also hard at work. A great opportunity to bond with our grandson.

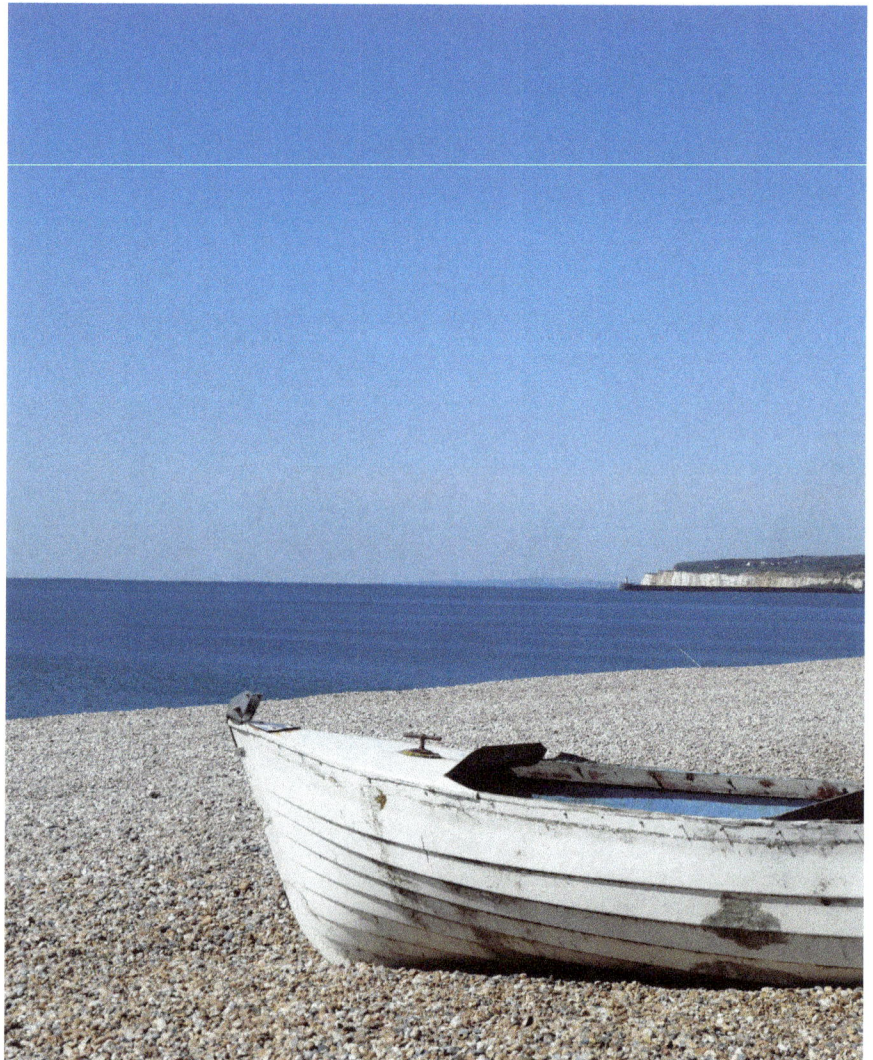

Our previous visit introduced us to the South Downs - spectacular walks and vistas. My image of the 'Long Man' demonstrates what's on their doorstep. Funnily enough, we'd previously visited this ancient icon. At the time of the wedding, my brother rented the Landmark Trust's brilliant renovation of the Priory in Wilmington for his family.

The earlier picture of the church is not of Selmeston, but Alfriston - a beautiful and special South Downs town. The amazing green adjacent to the church proved the perfect running venue for our Grandson – not sure if he's to be a sprinter or a long distance runner. Alfriston is well worth a visit – especially its amazing quirky bookshop, 'Much Ado Books'.

Authors & Books

The Kite Runner – Khaled Hosseini

I posed the question a few weeks back as how history would judge the withdrawal of US and UK troops from Afghanistan after 20 years. The fall of the Afghan Government to the Taliban may provide an answer. The 'Kite Runner' tells a story of the last time the Taliban were in power – let us hope the outcome is different this time.

Musicians & Popular Music

Sting – Fields of Gold

The image of the path through the corn fields on our way to dinner at the Sussex Ox - makes this an ideal choice. With Granny and Sting being Geordies, even more so. Takes Granny and **me** back to much younger days. (Eva Cassidy's rendition a favourite).

Composers & Classical Music

Benjamin Britten – Young Person's Guide to the Orchestra

Granny and I came late in life to classical music and would love to see you, our grandchildren, recognise the pleasure and succour this can bring to one's life. Hence the thinking behind this week's choice.

Fine Wine

2016 Rathfinny Estate, Blanc de Blanc Brut, East Sussex, England

When Granny and I were young, the idea of drinking English wines was not an option. As can be deduced from my image of the Rathfinny Wine Estate, a few miles from our son, daughter-in-law and grandson's new home - this has changed. So, fitting this week's wine comes from Rathfinny's impressive vineyard.

As ever, we're running out of space, and I should highlight the cliffs and promenade at Seaford – as per the imagery. Also a vineyard, this being on our son's doorstep and apt given his involvement in the wine trade. Plus the earlier picture highlighting our walk across the fields to the Sussex Ox. As you can see - plenty of excuses for return trips.

Granny

and I's first Australian adventure took place in 2008 and Byron Bay a destination on our itinerary, and a place we returned **to** a decade later. Byron Bay is reputed to be one of the most expensive places on Australia's East Coast to purchase a property. It's worth bearing in mind this is a fairly recent phenomena.

I'm reliably informed the Arakwal, Minjungbal and Widjabul peoples, the original indigenous inhabitants, lived on this coast for 20,000+ years before the 'white fella' arrived. Byron Bay has not always been the exciting, hip and trendy place we've come to know. The logging of cedar trees an early occupation (1830s), also at different times mining (gold, zinc, uranium and thorium), fishing and whaling further enterprises, farming and no doubt others.

Without taking time to look back it's easy to derive an incorrect view. This eminent tourist destination stems back to the hippy era of the 1960s and surfing: a time Granny and I can associate with. It remains a destination for surfers and backpackers - but has inevitably become more sophisticated and upmarket, our 10 year interval emphasising this.

This image epitomises Byron's trendy hippy culture and was taken from the balcony of an Aussie mate's apartment overlooking the beach, near the Surf Club. We've since learned Wicked Vans ran into trouble for their controversial slogans painted on their campers. We sadly quite enjoyed some of their more inane attempts at humour.

The slogan on the back bumper of this VW camper caught our eye: 'Patience is a Virtue'. The reason being my Mum (your great granny) was christened Patience – not the most apt of names, hence she only answered to Paddy.

The trees surrounding the van bring back fond memories of our first trip. Every evening there would be this amazing influx of Lorikeets (birds) coming into roost for the night – initially the noise incredibly high decibel, and suddenly it would cease. Next morning they'd depart as one - the swoosh of all those wings, tremendous.

Sadly on our visit 10 years later we found the Lorikeets no longer congregate in the trees by the Surf Club.

On our visits we've eaten at some great restaurants. One that sticks in my brain from our first trip was owned by Paul Hogan of 'Crocodile Dundee' fame – films Granny and I still like to revisit.
On our most recent trip the stand out restaurant was 'The Farm', a few miles out of town. Not enough room in the car, so two of us travelled via electric bike, mine a borrowed Vallkree - much more reminiscent of a vintage customised motorbike, very hip and cool.

Surfing and bogey boarding, as the picture demonstrates, is an extremely important pastime in Byron. Granny and I's attempts sadly nowhere near the required standard – we should have started at a much younger age. We loved having a go, but were put to shame by all the youngsters who'd obviously grown up on boards.

This image tells a different tale, and taken on our last visit. The coast had recently been hit by a cyclone which unfortunately removed much of the sand covering the beach – major damage had been inflicted and rocks exposed. Hence the helicopter, as someone was sadly killed while surfing and making contact with the rocks.

Cape Byron is the easternmost point of Mainland Australia and, we're told, a great viewing point to spot humpback whales, if you time your visit correctly. Unfortunately not whale season - but we saw Manta Rays, which appear extremely graceful. Byron has amazing walks both up and down the coast, with well-tended boardwalks and paths.

A walk embarked on and much enjoyed during both trips was out to the lighthouse, which involves a bit of elevation and provides spectacular views of the beach beyond – Wategos (1st image). We had a brilliant stroll way along this and then cutting back into Byron.

A further great walk was along Tallow beach towards the other end of Byron Bay.

As ever, lots to report but not enough space. It would be remiss not to explain our last visit involved meeting up with all our adopted Aussie mates from an earlier Sabbatical entry: Lake Como. The Impresario and his good wife's culinary skills as ever came to the fore. Making fresh pasta, spread out around the house – a new experience. The Spotty (Spotify) Game was possibly a tad rigged against ignorant Poms not au-fait with Oz musicians. As ever a great time had by all – and we look forward to travel opening up and being in a position to meet once again.

Authors & Books
The Last Rhinos – Lawrence Anthony

A book I found on a shelf in the house in Byron. Relates to a different continent (Africa) and the conservation of wildlife. A fascinating true story of one man's attempt to save a species and an insight to the horrors of war in the Congo. A book to make one think as to what can be achieved - as well as man's inhumanity to man and wildlife.

Musicians & Popular Music
Bobby McFerrin – Don't Worry, Be Happy

A song that chimes with Granny and me from 30+ years ago. Today's world often appears to worry too much? It's worth contemplating you can only alter things you're in a position to influence. 'No Worries' - a great Aussie saying.

Composers & Classical Music
Richard Addinsell – Warsaw Concerto

Byron deserves a composition encompassing the thrill and power of surfing the waves – also time to slow down and engage with friends and contemplate other things. The Warsaw Concerto combines both sentiments.

Fine Wine
2013 Leeuwin, Art Series Chardonnay, Margaret River, Australia

Too much fun and frivolity involved on our last Byron trip - so no comprehension of the wines we drank. Hence an Aussie wine consumed back in the UK - Christmas the same year. Granny and I've been to Leeuwin's impressive vineyard on a far too brief visit to the Margaret River. It's worth examining the reference as to the 'Arts Series'.

WEEK 34
MARCHMONT HOUSE
SCOTTISH BORDERS

When Granny and I embarked on compiling the Sabbatical in January of this year our ability to travel was severely restricted; we were in lockdown due to the COVID-19 pandemic. Our aim and intention was to invoke memories of amazing places, scenery, buildings, art, people, books, music and fine wine to raise both our and others spirits.

34 weeks on, COVID-19 is far from being eradicated. But it would appear we're learning to coexist with this insidious disease via vaccines and other medical advances. So once again we enjoy unrestricted freedom to travel within our own Country (as long as one is not infected) - but many restrictions remain in relation to foreign travel.

A consequence of travel restrictions and the pandemic has been the realisation and engagement with all that's on our doorstep. It was never our intention the Sabbatical should be constrained to far away and exotic locations.

Marchmont House is an excellent example and a place that's intrigued me for many years. Up to 6 weeks ago Granny and I had visited Marchmont just once in our entire lives – a hunt ball attended many years ago. Marchmont is a majestic and magnificent building 20 miles from our front door. As a Sabbatical entry it truly fulfils all the criteria.

An unusual aspect of this week's entry relates to 3 individuals, from 3 separate generations – none of whom I've ever met. But all have made, or are making, an indelible mark on Marchmont. It's not my intention to tell Marchmont's story – more able scholars having done so – but to record my interest in these 3 individuals.

An aspect of the Sabbatical has been the many crossovers and coincidences in relation to other entries, Marchmont having more than most. I never met Lady McEwen, who resided at Marchmont for many years - but growing up in the area, I was aware of the various tragedies that befell this illustrious family. It's not my intention to focus on these.

My interest in Lady McEwen relate to a couple of earlier Sabbatical entries: Monte Cassino, plus Paxton House and Chain Bridge. My Dad (your great grandpa) fought at Monte Cassino; he had ultimate respect for the Polish forces who eventually took the Monastery. Our choice of book in relation to Paxton House and Chain Bridge was 'Wojtek the Bear'. After the war, Wojtek resided across the river from us at Winfield: a camp for displaced Polish servicemen.

I'm led to believe Lord and Lady McEwen played an important role in helping the Polish people who'd so gallantly fought alongside the Allied forces during WW2 and subsequently lost their own homeland to Stalin's Communist Russia. To comprehend the enormity of this story, a lengthy tome worth reading is 'Trail of Hope' by Norman Davies.

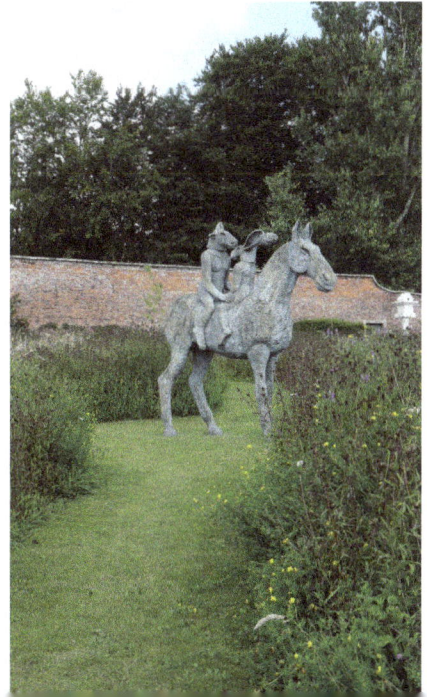

The 2nd person of interest relates to one of Lady McEwen's sons – Rory. I became aware of Rory and Eck McEwen around 10 years old, when the 2 brothers regularly appeared on the BBC 'Tonight' programme. Availability to music was so very different in those days, and this exotic pair were major stars and musicians – and of course local boys.

I was reacquainted with Rory McEwen a number of years ago when visiting Kew Gardens (London) and came across the retrospective exhibition of his amazing botanical paintings. The talent of this man as both a painter and musician is worth exploring. But the aspect that most attracts me relates to his humanity and approach to his fellow man. Something we could all learn a great deal – an extract from a tribute by his friend Alistair Reid:

"Rory was always to me a very extraordinary creature, just in himself. When I think of Rory now I think of that incredible grace of his in every respect, grace not only in a physical sense, but in his great care and concern in talking to anybody. He listened with maximum attention to people and took them in and reacted. This was the most unforgettable and admirable thing about him, his absolute total attention to anybody, no matter who they were".

The 3rd person is the current custodian of Marchmont – Hugo Burge. For the last decade the Burge family, via Marchmont Farms Ltd., have been restoring the house and grounds to their former glory. Granny and I can attest to this, having been on a sculpture tour 6 weeks ago. The foresight, aim and ambition of this project is astounding, and we salute all involved.

I should point out a further Sabbatical coincidence being an Anthony Gormley sculpture on Marchmont's roof. Rather than attempting to explain more I will leave the imagery to provide an insight – and highly recommend a visit.

Authors & Books
The Colours of Reality – Rory McEwen

I bought this book after our visit to Kew Gardens (London). It brings to life the skill of Rory the artist and the musician, but also a background to an extraordinary life and an insight to the past glamour of Marchmont.

Musicians & Popular Music
Eric Burdon and the Animals – House of the Rising Sun

A prominent Geordie band of the 60s, which takes Granny and I back many years. Relevant in that Eric Burdon and many other future stars were heavily influenced by Rory McEwen and his love of Leadbelly and the 12 string guitar.

Composers & Classical Music
Pietro Mascagni – Intermezzo from Cavalleria Rusticana

A piece of music we both enjoy, and it lends itself to Marchmont and its exciting future. An amazing and inspiring venue and project which we're lucky enough to have right here on our doorstep – the music chimes with this.

Fine Wine
2001 Domaine Pavelot, Savigny-les-Beaune La Dominode Premier Cru, Burgundy, France (magnum)

Marchmont deserves a fine wine, one with character and in-line with the aims of the Sabbatical. A wine consumed with friends and family at the Laxford (one of our favourite places) over a decade ago.

WEEK 35
MELBOURNE & YARRRA VALLEY AUSTRALIA

We're

18 months on from the first UK COVID-19 lockdown and, for all intents and purposes, the Government in Westminster appears to have decided society should return to normality with minimal political intervention, the onus placed on the individual – plus a belief vaccines will prove the panacea.

With 35 to 40 thousand infections being recorded on a daily basis, hospital numbers at 8,000 and 100 to 200 hundred COVID related deaths recorded on a daily basis, it's fascinating to liaise with our Aussie mates in Melbourne, who are enduring their 6th lockdown – their figures infinitesimal compared to the UK.

It will be interesting to see how history records the outcome. Can the pandemic be controlled by locking countries down, or do we learn to live with the disease to enable societies to thrive and prosper? Many support one or other side of the argument. I don't profess to have a crystal ball - but doubt locking down will prove a long-term answer.

Last Christmas Granny and I should have been in Melbourne to celebrate a great friend's 70th birthday. For obvious reasons this did not take place – and is currently on hold. Melbourne is our favourite Oz City - so our thoughts and best wishes go to all the many Aussie mates we've been lucky enough to team up with on our various visits.

We look forward to meeting up once again in person - whenever that should prove possible.

Melbourne is a favourite of ours for a number of reasons: the people, food, facilities and events, to name just a few. The first image is the view from the end of the pier at St Kilda looking back over the city – a great place for a walk, a drink, and to see the miniature penguin colony that's made this home. Well worth a visit, and takes one back in time.

The second image demonstrates the view from our digs while staying in Melbourne – looking out over the bay from Beaconsfield Parade, a short walk or cycle ride from St Kilda. One of the great things about Melbourne is the plethora of cycle-ways, to bike the 5km from Beaconsfield Parade to the heart of the City; a doddle.

This a picture of a house in Middle Park, behind Beaconsfield Parade, an area I find fascinating. The houses date back to Victorian times, a forerunner to our prefabs. It's a clever piece of urban development involving big wide tree-lined streets. What originally was a working class area is now an incredibly trendy and expensive part of town. The sympathetic and ingenious development of the original properties is clever, and appeals to my interest in design.

Our first visit to Melbourne coincided with the Australian Grand Prix, which involves a street circuit built around Albert Park, which abuts Middle Park. Granny and I were lucky enough to go and see this in action. Unlike the British Grand Prix, no massive traffic snarl-ups; just a short stroll from where we were staying. Unfortunately no stunning imagery, as my attempts to capture a F1 car at 200mph were not worthy of the Sabbatical.

A further memorable event on another visit to Melbourne, right up Granny's street - the Australian Open Tennis. Granny is an avid or possibly obsessed tennis aficionado, so attending the championships was a major highlight. Especially when this involved watching her tennis idol Sir Andy Murray playing Novak Djokovic in a thrilling 5 set match, taking almost 5 hours. Sadly Andy was beaten at the very death – but Granny loved the experience.

I've mentioned cities are not my natural home – but Melbourne, with its bay-side promenade, pier, cycle-ways, the river, plus its spectacular Botanical Gardens provide me with the open spaces and tranquillity I enjoy. This image from the Botanical Gardens represents a tiny aspect of this amazing facility. We've visited on numerous occasions.

On our way to the Gardens on the last trip we visited the impressive Shrine of Remeberance, commemorating Australia's involvement in World conflicts, which provides amazing vistas of Melbourne from the roof terrace. It's apt and poignant to record the involvement and sacrifice Australians have made on behalf of others.

As ever, space is short, but I must mention the Yarra Valley, famous for its production of wines. Granny and I've made 2 trips to this fascinating region. The image relates to Yering Station (1st visit) – great location, wine and food. On our last trip we'd lunch and a wine tasting with friends at Moet Chandon's winery: excellent. Highly recommend a visit.

Authors & Books

Tattooist of Auschwitz – Heather Morris

A Melbourne resident who survived Auschwitz. His story not told until the latter years of his life – a harrowing but true tale. The sequel, 'Cilka's Journey', recounts a fellow inmate's subsequent misfortune – from Auschwitz to a Soviet Gulag. Two books providing insight to Nazi and Soviet horrors and atrocities.

Musicians & Popular Music

Sade – Smooth Operator

Sade and 'Smooth Operator' transports Granny and me back many years and memorable times (played on the latest portable gadget – a Sony Walkman). The title fits the bill, and an appropriate reference to a day out in a fancy car - as recorded in this week's fine wine entry.

Composers & Classical Music

Vittorio Monti – Czardas

The tempo follows this week's Sabbatical tale perfectly – from a serious debate about the consequences of the pandemic, to much more enjoyable memories. Also appropriate in melody references the 'Tattooist of Auschwitz'.

Fine Wine

(unknown vintage) Yering Station, Chardonay and Pinot Noir, Melba Highway, Yarra Glen, Victoria, Australia

Memories of a special day out with Granny, having been lent a very smart 3.0 litre convertible BMW. We looked the business. A great setting with a mixture of original vineyard buildings and a new contemporary restaurant. A great lunch - but reserved the wine for consumption with very special friends on our return to Melbourne.

WEEK 36
EILEAN DARACH &
BENMORE ESTATE
SCOTLAND

A recent question asked in relation to compiling the Sabbatical: What's the most difficult element – the places, words, photography, books, composers, popular music, or fine wines?

This week's entry demonstrates photography, and capturing the relevant images the most problematic. If you're in a location for a limited period of time, and at a distance, plus the weather not playing ball – there's an issue. Eilean Darach and Benmore Estates highlight the conundrum.

We've just returned from a few days at Eilean Darach and fishing the Gruinard. The first image demonstrates the weather being perfect for photography – taken on arrival, late afternoon. The second an hour later. The following 3 days as demonstrated by the third image, overcast and far from ideal for photography.

Benmore Estate on the island of Mull involved the opposite conditions: the weather overcast or wet until the morning of departure, and time severely restricted as we'd a ferry to catch. So the answer - photography.

Eilean Darach and Benmore highlight an important aspect of the Sabbatical. Week 30 (Glen Lyon and Loch Tay) incorporated a memorable celebration, my 65th birthday. Mentioned within this, a speech I gave to 24 great mates; a short extract epitomises what the Sabbatical aims to promote:

Great friends of ours use a phrase to which I totally concur: 'Families that play together, stay together' – to which I would like to add the word 'Friends'. If you would be upstanding I would like to propose a toast: 'To Family and Friends, and to lots more playing together in the years ahead'.

The recent few days in Eilean Darach, our third such visit, highlights being with friends in a beautiful setting. COVID-19 and lockdowns have emphasised the importance of such occasions. A big thank you to our hosts – we loved it.

Before moving onto Benmore and Mull: I should explain not all aspects of our vacations prove idyllic and go to plan. Unfortunately on this last trip, the rain gods not on our side; the river on its bones, almost an all-time low. Hence Granny and I failed dismally in relation to catching a salmon.

A further illustration of things going awry: I took Granny 5 miles up the Little Gruinard to Fionn Loch. As the prior image demonstrates, the track basic and progress slow. On arrival we went to fish a couple of runs; I'd forgotten the midge nets, so left Granny to fish and went to retrieve these. On my return Granny appeared to be doing the Highland fling, or performing a lively semaphore SOS - having been eaten alive by midges. I was far from popular.

This image illustrates Tobermory on the island of Mull and brings back happy memories of a gathering to celebrate a great friend's 60th birthday (2019); this the family who introduced us to the phrase 'Families that play together, stay together'. Three generations involved in the festivities, and proved a memorable and highly enjoyable sojourn.

Our stay at Knock House, Benmore Estate, unfortunately lacks the imagery to demonstrate what an amazing location this is (see earlier explanation). Here are some hastily snatched shots prior to boarding the ferry. This next image shows Loch Benmore, a stunning destination. Granny and I would love to return - the scenery spectacular.

Our 4 days on Mull were full of contrasting escapades: salmon fishing on the river, trout fishing on Loch Benmore, a trip to the island of Inch Kenneth, a special dinner to celebrate a 60th birthday – to highlight just a few.

Our day on Inch Kenneth deserves mention: we travelled in style aboard the Lady Benmore and hiked around the island, before a damp but enjoyable barbecue on the beach. I'd an amazing sense of deja vu on coming across Inch Kenneth House.

Research on my return established this had been the former home of Unity Mitford – notorious for her love affair with Adolf Hitler and Fascism. A tale of warped allegiances to Nazism, Fascism and Anti-Semitism!

This final image taken from the ferry on the way back to Oban. Our brief encounter with Mull has whet Granny and my appetite for more, and a return trip definitely on the cards.

Authors & Books
Born Survivors – Wendy Holden

A true story recounting babies born within Nazi death camps and survived – later in life, to meet in person. The story of Unity Mitford highlights the dangers of fanaticism, bigotry and ignorance. Today's world of social media and extreme polarised blinkered views – to my mind, requires to be treated with a healthy dollop of scepticism.

Musicians & Popular Music
Simon and Garfunkel – Sounds of Silence

From Granny and my teenage years, but fitting in relation to the Highlands and Islands – the sense of space, peace and quiet is magical and makes trips to the likes of Eilean Darach and Benmore special.

Composers & Classical Music
Dvorak – Symphony No 9 (New World)

Salmon fishing trips involve expectation – wake in the morning to rain, the river rising and the possibility of fish on the move. On the other hand, sun shining and little or no water – provides an opportunity to take in the amazing scenery of the Highlands and Islands of Scotland. Dvorak manages to encapsulate the vagaries of such.

Fine Wine
2007 Domaine Hubert Lamy, Santenay Premier Cru Clos des Gravieres, Burgundy, France

No record of the wine consumed at Knock House (Benmore) – but the event deserves a 'Premier Cru', hence a wine from my late brother-in-law's cellar, drunk with family in the Lakes, Christmas 2014. Mull a special and memorable birthday celebration of a great friend - involving 3 generations; something Granny and I love to see and encourage.

WEEK 37

CASTELL D'EMPORDA, AIGUA BLAVA & GIRONA SPAIN

This has been a week of firsts (in relation to the onset of the Pandemic): Granny and I's **first** flight, **first** stay in a hotel, **first** time on foreign soil, **first** wedding, and **first** stay in an ancient Castle. It's been quite a week.

Navigating the rules, regulations and different protocols proved an interesting experience. Living on the English side of the Border and flying from Edinburgh to Girona added complexities and issues, which at times **were** hard to fathom. COVID-19 Passports, test kits for our return, tests before returning, locator forms ref exiting and entering countries – all a new and novel experience. Luckily I passed the exam, having got Granny and I out of the UK and back.

As much of my working life was spent living in hotels, it's not my first choice when going on holiday. But when staying in a Castle dating back to 1301, and the view from our room is as per the image, I can be persuaded. Castell D'Emporda is an impressive sight atop a high vantage point. Delighted to report facilities have advanced over the 700 years. The contemporary interventions appealed to my passion ref design, and I compliment the Dutch owners.

Our accommodation proved excellent and the staff looking after us, first class and delightful. A novelty to stay in a hotel and be waited on hand and foot – particularly after lockdowns and all the various restrictions endured since March 2020. Makes one feel a little like a naughty schoolboy escaping the clutches of authority – the behaviour of certain members of our party could well be assimilated with such. A great time had by all.

157

I've alluded to the fact at various times within the Sabbatical my preference for wide-open uninhabited spaces. A visit to the Costa Brava would normally fill me full of trepidation, conjuring up a picture of mass humanity. A brief visit to the coast prior to the wedding proved benefits can emanate from a pandemic – the town and beach all but deserted. Tough for all the businesses - but my kind of place, and likely to be a one-off experience.

The wedding we attended, being that of the son of great mates who've been party to various Sabbatical entries, COVID-19 discouraged many of our age group from attending. But as members of the 'A-Team' we felt we'd little alternative other than to support our friends. The event having been cancelled on 2 prior occasions due to the Pandemic, we wouldn't have missed it for the world – a terrific occasion.

The younger generation were obviously far less inhibited about flying abroad and partying - far outnumbering their parents' generation. Possibly a lesson in there for those organising such events in the future? It would be remiss not to point out 4 generations attended the wedding. The bride and groom had grannies and grandpas (in their 80s and 90s) attending – a valiant effort. A number of the young married couples also with babies in tow.

Granny and I have extolled the virtues of multi-generational interaction – this is a perfect example.

The wedding ceremony took place in a beautiful olive grove – the microphone picking up the thunder, adding a unique atmosphere to proceedings. The bride and groom looked amazing, and the ceremony completed and vows exchanged ahead of the gathering storm. The rain in no way

dampened events, with everyone undercover by that stage. A memorable time had by all, and my dance lesson with a beautiful ballerina a highlight.

We've attracted an eclectic group of friends over the years, as you may have gathered – one of whom collects bricks! Hence our trip to the Brick Museum at the nearby town of La Bisbal. A much more enlightening experience than envisaged. The town had in years gone by been a major centre for the manufacture of bricks and ceramics. This week's popular music choice had been reserved for a further Sabbatical entry – but apt in being reallocated.

Our return flight provided a full day - of which we made good use. A trip to Pals, a fascinating walled settlement even more ancient than our Castle. Then a trip down memory lane for lunch. 35 plus years ago, Granny and I holidayed in Aigua Blava with good mates (Sue and Denbigh), 2 years in a row. Memorable nights in Harry's bar. Sadly they're no longer with us – one succumbed to a brain tumour, the other a heart attack. Great friends who are missed.

Our final destination: Girona. An impressive city, and one we'd like to return to. The river, Cathedral, narrow clean cobbled streets, impressive squares - worth a visit. Not so sure about traipsing all over in search of Massimo Duti?

Authors & Books
**I'll See Myself Out, Thank You –
Colin Brewer**

The subject may appear morbid, but lunch in Aigua Blava saw Granny and I reminiscing as to a great friend with an inoperable brain tumour and the courage to refuse treatment to live her last 6 months in style – makes this worth examining.

Musicians & Popular Music
**Pink Floyd –
Another Brick In The Wall**

A visit to the Brick Museum in La Bisbal makes this the natural choice. Pink Floyd very much part of Granny and my younger days. Originally earmarked for an entry involving my attitude and approach to education – which came to a premature end!

Composers & Classical Music
Arturo Marquez – Danzon No 2

This composition emphasises the build-up and frenetic nature of our 4 nights in Spain. So much packed into a limited period of time and involving so many highlights and contrasts – hence a perfect choice.

Fine Wine
**2016 Descendientes de J. Palacios
Moncerbal, Bierzo, Spain**

Granny's preference in wine relates to white, a crisp dry French Burgundy high on the list. When it comes to Red wine, she's partial to a mellow Spanish Crianza. Various Rioja Crianza's consumed over this last week – but no record. So a wine consumed with family and friends on a previous occasion.

WEEK 38
BEAMISH & WOODHORN MUSEUMS ENGLAND

The Sabbatical aims to conjure up memories via places, imagery, words, music, books, and fine wine. This week's entry takes Granny and **me** back many, many years and invokes a panoply of reminiscences in relation to times gone by. It fascinates us **that** so much has changed during our lifetime, much **of** which is alien in the realms of today's world.

Believe it or not, we'd no computers, mobile phones, or social media when attending school. A great example relates to calculators – our equivalent, a slide rule, which neither of us ever managed to master. A very different and much simpler way of life: we were expected to entertain ourselves, the great outdoors our playground.

Granny and I wondered how we could impart the contrast of childhood in the 1950s to the lives you, our grandchildren, lead today. Hence the inclusion of Beamish and Woodhorn Museums, which provide an insight into the immense change that's taken place during the time we've been on this planet. A trip down memory lane.

Beamish Museum is like stepping back in time; way beyond us, but much of what's portrayed in use and vogue when we were little. The people who decided, 50 years ago, the past should not be lost, deserve praise. Their foresight in creating an amazing repository of life in relation to bygone times resonates with Granny and me.

Picking half a dozen images to demonstrate the two museums does them a total injustice – our aim is to provide a mere taste and insight as to what these entail. Beamish Museum covers 350 acres and incorporates the most amazing array of buildings and artefacts aimed at highlighting the North East's industrial and social heritage.

Beamish incorporates buildings original to the area and others which have been painstakingly dismantled, transported and rebuilt on the site, in addition to various replica structures and dwellings of their time. Furthermore, the collection of artefacts associated with each of these paints an amazing picture of life in relation to their context and period.

The eras Beamish encompass date back to the 1820s, with aspects even older. Their 1820s landscape includes Pockerley Old Hall, farmstead and garden; Pockerley Waggonway and Great Shed highlighting the transformation coal wrought on the area and Britain's economy. St Helens Church and Quilters Cottage 2 more recent additions.

The 1900s town involves re-enactments of what took place at that time – as demonstrated by the image of Jubilee Confectioners. Within the town you can see and visit the bakery, chemist and photographers, garage, Co-op store, the music teachers, dentists, solicitor's office, pub, stables, printers and newspaper branch office, sweet shop, bank, masonic hall, fair, park, and bandstand – to name just a few. An amazing insight to the early 1900s.

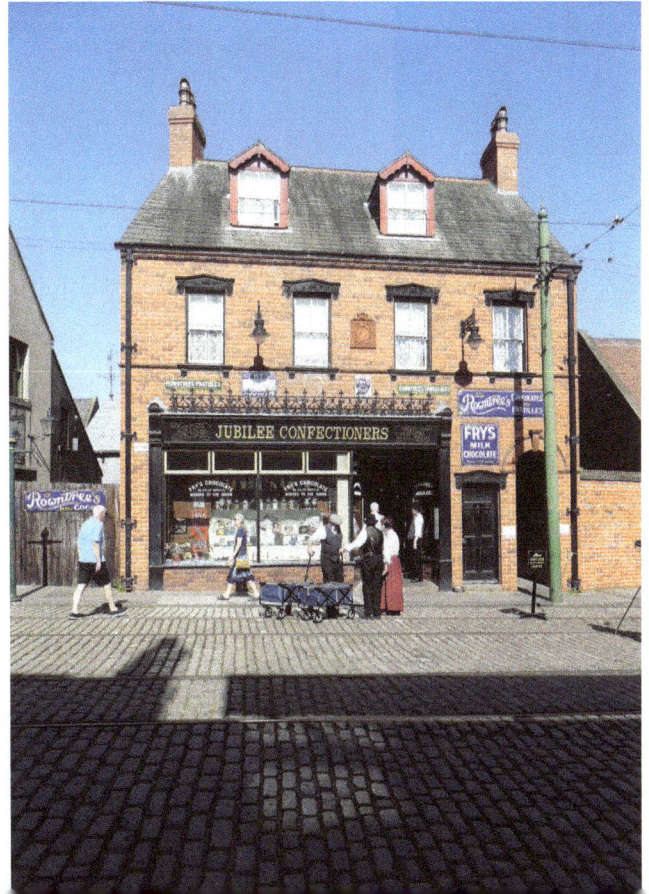

The development of transport from the 1820s onwards - a delight. From pack horses to carriages, waggonways to steam trains, trams to trolleybuses, bicycles to motorbikes, cars and buses. Beamish has its own railway station, tramway and road network. A fascinating insight to the development of transport across 2 centuries.

The 1900s pit village provides an insight to life in a mining community: its colliery and drift mine, rail yard, engine shed, stables for the pit ponies, smithy, terraced houses, chapel, school, band hall, fish and chip shop – and much more. It aptly demonstrates how life has altered in a very short period of time.

The 1940s farm highlights a period prior to and during the 2nd World War: the farmhouse, cottages, netty, farm steading and buildings, machinery, animals and fields all highlight the massive advances in agriculture since that time. Beamish incorporates and evokes so much more - the only way to truly appreciate this is to pay them a visit.

The final image shows Woodhorn Colliery outside Ashington (Northumberland) – this now a museum. Ashington at one stage was the largest mining community in Europe, if not the World. Woodhorn Museum highlights the importance and influence coal and mining had on the local and national economy. A further museum to expand your knowledge.

At sixteen I attended Ashington Technical College, living in digs with a mining family - an abiding memory of the outside 'Lavvy' in the backyard. I failed my exams due to lack of engagement, but learned much about people and society. I've fond memories of that time, and recognise the profound influence this had on my outlook to life.

Authors & Books

Billy Elliot – Melvin Burgess

I saw the film and musical prior to reading the book – takes me way back to my teens and living in a mining community. A great story and lesson: follow your passion and be willing to step outside the norm to succeed.

Musicians & Popular Music

Barbara Streisand – The Way We Were

A great favourite of Granny's, and relates to memories and times gone by – relevant to Beamish and Woodhorn.

Composers & Classical Music

Dimitri Shostakovitch – Waltz No2 from Jazz Suite No 2

Shostakovitch's Jazz Suite No 2 somehow encompasses a way of life that's vastly changed - class and social politics major topics of debate in our younger days. Shostakovitch's life provides an interesting insight to contrasting thoughts and views as at that time - 'The Noise of Time' by Julian Barnes may help to explain.

Fine Wine

2020 Domaine Simon Bize et Fils, Bourgogne Blanc Les Perrieres, Burgundy, France

Invoking memories and looking back on good times – very much the backbone of the Sabbatical. This a wine we drank in New York with our son and daughter-in-law (your Mum and Dad, or uncle and aunt) while they worked in NY.

WEEK 39
PALM COVE & DAINTREE RAINFOREST AUSTRALIA

It's 39 weeks (New Year 2021 and the 2nd Lockdown) since we embarked on our Sabbatical journey with a letter to the grandchildren. The image utilised for this taken during our 1st Australian trip, shot from a small aeroplane when travelling from Palm Cove to the World Heritage-listed Daintree Rainforest. A very special and memorable 10 days.

We flew into Cairns and travelled onwards to our hotel in Palm Cove – directly on the beach. We fondly remember dinners on the terrace looking over the ocean, the moon shimmering through the palm trees – very romantic. The hotel designed in a relaxed traditional colonial style, the staff delightful and attentive. Granny and I were in 7th heaven.

This 2nd image demonstrates the amazing beach, and almost totally deserted – secluded walks part of our daily routine. Sadly swimming restricted to a netted-off area in the sea, as it was prime Stinger season, these being jellyfish packing a powerful punch. Slightly perturbed to find the previous week, a salt water croc had inhabited our protected swimming enclosure – luckily not in residence during our visit.

A highlight of our stay in Palm Cove involved an excursion from Cairns out to the Great Barrier Reef – a spectacular experience and, if you get the opportunity, well worth doing. Granny was reluctant to snorkel in the wide-open expanse of the ocean. On the other hand I relished the experience, the breadth and variety of fish species being truly spectacular.

An amusing escapade worth telling relates to our journey back from Cairns to Palm Cove: this involved a shopping trip to find suitable luggage for our flight in a light aircraft to Bloomfield Lodge. The stipulation being the maximum size (of the one case allowance per person) was 30cm x 50cm x 15cm. When Granny saw how small these were, and how little she could take, she had a hairy canary and stormed out of the shop!

This picture shows Weary Bay, Cape Tribulation – so named when Captain James Cook's ship the HM Barque Endeavour ran aground in this location (10th of June 1770) and required extensive repairs before resuming their voyage. I will leave you to explore the history of Cook and the Endeavour – much more able scholars having done so.

Our plane landed at an extremely rudimentary landing strip off to the right, deep in the forest. An interesting aspect was the low-level run to chase away the kangaroos before the plane set down. We then travelled via a 4WD vehicle to the river and onward by boat to the lodge - which was totally hidden by the dense forest on the opposite bank.

A trip with a wide variety of first time experiences. Shane, our resident 'Crocodile Dundee' - note the bare feet while showing us the delights of the rainforest. As you can see, a snake was investigated. We also foraged for bush tucker, which involved eating live ants which tasted of lemon – an alternative ingredient for gin and tonic. Witchetty grubs – large, live, white wood-eating larvae – were also on the menu. Refused by all, bar me; they tasted surprisingly good.

Shane, Granny and I went in a small boat up the Bloomfield River via the mangrove swamps to explore. The bird life spectacular and colourful, as was our encounter with this handsome fella – a very large saltwater crocodile. We learned (via Shane) it's not a good idea to camp near the river for more than one night – a German tourist having recently found this out the hard way, ending up as dinner for a large croc.

A further experience during our visit related to the heavens opening and a monumental rain storm. This happened in the middle of the night, and the noise on the corrugated iron roof of our plush shack was epic. We got up and sat on the veranda and took in this extraordinary sight. The following morning, little evidence of the large volumes of water.

The final image shows my favourite spot – I consumed a few 'Stubbies' while fishing from the end of the jetty, shaded from the heat of the sun. I managed to catch a 2 foot reef shark – extracting the hook and returning the fish without losing my fingers proved an interesting experience. Bloomfield Lodge was a great and memorable adventure; the staff who looked after us and the various people we met made our stay very special.

A fascinating snippet turned up in this week's Telegraph: 'The World Heritage-listed Daintree Rainforest was handed back to its traditional Aboriginal owners in a deal signed with an Australian state government yesterday'. Daintree being one of the oldest rainforests in the World, more than 130 million years old.

Authors & Books
Taboo – Kim Scott

Not an easy or comfortable read - but an insight to a past massacre of indigenous people and moving onto more contemporary times and issues. Another recommendation from my favourite bookshop in Albert Park (Melbourne).

Musicians & Popular Music
Phil Collins – In The Air Tonight

Phil Collins is a long-time favourite of both Granny and me, and takes us way back in time. Not so sure the lyrics fit this week's entry - but the ethereal nature of the composition has an empathy with Daintree and the rainforest.

Composers & Classical Music
Claude Debussy – Arabesque No 1

Debussy's music resonates with Granny and me: Arabesque No 1 chimes with a very special 10 days, in 2 memorable places. An apt time to look back and reminisce as we celebrate our 44th wedding anniversary this week.

Fine Wine
2007 Mazis Chambertin, Grand Cru, Cuvee Madeleine Collignon, Eleve Par Albert Bichot, Hospices de Beaune, Burgundy, France

We've no remembrance of wines drunk at either Palm Cove or Bloomfield Lodge. So a very special wine supplied by and consumed with the Laird and his good Lady - while salmon fishing with friends (this last week) in Glenlyon.

WEEK 40
KELSO, DRYBRUGH, MELROSE, & JEDBURGH
SCOTTISH BORDERS

The Sabbatical is 75% complete, and nobody more surprised than **me** that we've managed to reach this milestone. There was an element of trepidation when embarking on our project – but so far so good. The intention from the start being the Sabbatical should encompass much more than faraway destinations – hence a return to our locality.

Having been away fishing for a week and shut off from the media, it's interesting to see headlines relating to the various crisis purportedly about to engulf our country: climate change, COVID-19, lack of fuel, energy prices, NHS, scarcity of drivers, protesters blocking motorways, mental health issues, inflation, food shortages, education, etc.

These possibly pale into insignificance when looking at the life and times of the Abbeys of Kelso, Dryburgh, Melrose and Jedburgh. Wars and the Protestant reformation, amongst others, devastated these once-wealthy and thriving institutions. Much has been written about these, and I leave you to examine their fascinating rise and fall.

I had my own minor crisis when attempting to photograph these intriguing buildings, all recently fenced off due to the danger of falling masonry. Hence the opportunity for exciting imagery greatly curtailed - for now.

The scale and influence these institutions must have had in their heyday is quite staggering, and all within a 20 mile radius of one another. Who needs to travel to far-flung places when epic edifices such as these are on your doorstep? A visit to the Abbeys brought back memories for Granny and me, covering a great number of years.

The first image highlights Kelso Abbey, a town with which we've had involvement. This second image shows the magnificent Floors Castle (Kelso), a place Granny and I have visited on a number of occasions. Granny at one time ran a franchise of her business within Sunlaw's Hotel, and her landlord resided at Floors (the late Duke of Roxburgh). His greeting always began with 'I see you've brought your minder along' (me). He proved a fair man to deal with.

The second Abbey on our tour, that of Dryburgh: possibly the most intriguing, as it appears to be set in the middle of nowhere and possesses a fascinating atmosphere and charm. Within the Sabbatical we've visited various magnificent buildings and gardens that were, or are, made possible by enterprising individuals. Two which stand out for Granny and me are Cragside and Marchmont. Dryburgh deserves to be included in this category and straddles the timelines.

David Erskine, the 11th Earl of Buchan, purchased Dryburgh House in 1890 and set about creating an amazing landscape encompassing the Abbey ruins and grounds. During the 40 years he lived at Dryburgh, he set about developing the gardens and parklands, this involved various interesting follies, but at the same time preserved the ruins of the great Abbey. Imagine how little may have remained if Erskine had not set about his Herculean task.

A couple of interesting facts relating to Dryburgh Abbey: Sir Walter Scott, one of the most influential writers of the 1800s, is buried within the ruins. Also Field Marshal Earl Douglas Haig of Bemersyde - viewed by some as a controversial, but nonetheless an important military figure in relation to the 1st World War.

This next image ties Kelso, Dryburgh and Melrose to Granny and me. You'll have previously ascertained from the Sabbatical, salmon fishing is a pastime we much enjoy. My picture highlights an extremely competent young fisher lady and her attentive ghillie. Over the years we've fished both the Teviot and Tweed, at Kelso, Dryburgh, and Melrose. I fished many times with your great grandpa at Middle Pavilion (Melrose) – involved lunch at Bert's Hotel.

Talking of Melrose, here is the Abbey: once again, the scale of what this enterprise must once have encompassed is truly immense and worth investigating. Melrose is a small Border town, and well worth a visit. One of its major claims to fame, other than the Abbey and fishing, relates to rugby as they are hosts of the World-famous Melrose 7s.

Our final destination is Jedburgh Abbey, once again on a grand scale; unfortunately our access to illustrate this greatly hindered. My major claim to fame in relation to Jedburgh: me and my siblings learned to swim there (many years ago) – our instructor a Mr Motion, this the only indoor pool in our area at that time. Fond memories of my Mum (your great granny) racing the Kelso Flyer (a steam train) in her Morris 1000 – a time before Mr Beeching.

Authors & Books
Pillars of the Earth – Ken Follett

A fascinating book providing an intriguing insight to the building of a Cathedral in 12th century Britain. In addition, an awareness of life and society as at that time. Relates well to the Abbeys of Kelso, Dryburgh, Melrose, and Jedburgh. On your tour call into 'The Main Street Trading Company' (St Boswells) – a bookshop with a difference.

Musicians & Popular Music
Moody Blues – Knights In White Satin

Knights in White Satin dates back to 1967 and **first** appeared on the Moody Blues album 'Days of Future Past'. An influential album from Granny and my teenage years – whose title fits well with the history of our Border Abbeys.

Composers & Classical Music
JS Bach – Toccata and Fugue in D Minor

A monumental piece of music thundered out on a vast organ – totally fits the bill in relation to the scale, power and influence these institutions held during their particular time in history. Totally apt in relation to this week's entry.

Fine Wine
1995 Château Ducru Beaucaillou, 2eme Cru Classe, Saint-Julien, Bordeaux, France

Salmon played a major part in the finances and diet of the Great Abbeys of the Borders. So fitting this week's wine consumed at a Salmon Club Dinner in 2016. Also marks the sad loss of one of our members during this last week.

WEEK 41
LAUNCESTON & CRADLE MOUNTAIN TASMANIA

An element of escapism is not a bad thing given the current climate of increasing cases of COVID-19, as well as a spike in hospital admissions. One has to ask whether our government's policy of learning to live with the disease will prove correct, or whether they'll be forced to reintroduce aspects of the restrictions they've dispensed with.

A trip down memory lane to Tasmania is a worthwhile diversion from a constant barrage of negativity.

Granny and I spent a delightful week in Tasmania with 4 great mates from the UK who happened to be touring Australia at the same time as us. A recommendation from Aussie friends sent us to Launceston in North Tasmania before travelling to Cradle Mountain. This is a favourite image taken way up the mountain trail – spectacular views.

Our second night involved a trip to one of Launceston's top restaurants, 'Stillwater', housed in an historic mill building. The dinner memorable in a number of ways: the food, service and bonhomie truly excellent, but the bill eye-watering for us poor Poms, the exchange rate at that time stacked massively against us (Oz $ and the £).

Peppers Seaport Launceston proved an ideal launch pad for our adventures – the accommodation great and the staff delightful. As the image demonstrates, our balconies looking out over the river. A tourist boat on the Tamar River encompassing the Cataract Gorge and a spectacular walk (in the rain) provided a useful insight to Launceston.

Sadly the weather not conducive to photography, hence no imagery.

Launceston's colonial history goes back to the early 1800s; the Queen Victoria Museum provides an ideal opportunity to investigate further. I also much enjoyed my visit to the Design Centre Tasmania. A further worthwhile trip involved a brief visit to the James Boag Brewery, founded by a Scot some 160 years ago. Launceston incorporates some impressive buildings dating back to Victorian times. I leave you to find out more.

A further adventure involved a visit to a Tasmanian Vineyard – Pipers Brook Winery. Once again the weather not playing ball, so no imagery – but we'd a great lunch and an entertaining wine tasting, and also purchased some wine.

Our drive from Launceston to Cradle Mountain provided an insight to Tasmania's wealth of forestry and why its landscape is so well suited to Hydropower. An unusual agricultural experience (1 of our party being a farmer) related to an unorthodox crop – turned out to be cannabis, and legally grown for the pharmaceutical industry.

Tasmania's wildlife proved interesting as demonstrated by this image of a Wallaby and her young Joey. We stayed in some wooden lodges adjacent to Cradle Mountain, and the Wallabies proved extremely friendly, almost knocking on our door. We also learned about the ferocious Tasmanian Devils and the cancer endangering the species.

This week's opening image from our trek to Cradle Mountain. The route varied from excellent to shabby chic, and onwards to downright dangerous. Not sure Granny and I were totally prepared for the extent of our day out, which began with a walk around a number of lakes and subsequently gaining elevation - at one point involving ropes.

The final 2 images provide the contrast in conditions: an amazing, well-maintained and manicured boardwalk looking into the distance, enticing us onwards to a scree track which at times was difficult to identify and extremely steep. The views spectacular, and well worth expending so much energy and effort ascending the Mountain.

Our short visit to Tasmania a number of years ago has whetted our appetite for more. One couple from our party went on to Hobart and visited the world famous museum of old and new art (MONA) – this now on our bucket list.

Tasmania, originally called Van Diemens Land, is notorious for the part it played as a British Penal Colony. 75,000 convicts served their sentence in Van Diemens Land, 67,000 of these transported from British and Irish ports. For those who think life in today's society is tough – delve into the brutal conditions and harsh environment these poor souls endured. Many of their crimes unlikely to invoke a police caution in our modern world.

Authors & Books
**Closing Hells Gates –
Hamish Maxwell-Stewart**

Gain an insight to the life and depravity involved in a penal colony in the early 1800s and contrast this with the freedoms of our modern day UK society. To comprehend this, read Hamish Maxwell-Stewart's tale of Macquarie Harbour, designed to house those convicts who failed to conform. Concludes with a wonderful sting in the tale.

Musicians & Popular Music
Carole King – So Far Away (Harvest)

Carole King takes me back to my teenage years and the angst of those times. The title fits well with COVID-19 and the thought of escaping to the other side of the World – hopefully something that'll become possible once more.

Composers & Classical Music
Edvard Grieg – Peer Gynt Suite No 1

Grieg's composition apt in a variety of ways in relation to this week's Sabbatical entry. Granny and I attended our nephew's wedding at the weekend and my dinner partner (Gillie, the organist, played this prior to the service commencing). Also the theme and direction of the music fits our varied Tasmanian itinerary beautifully.

Fine Wine
2014/2018 Vasse Felix, Heytesbury, Margaret River, Chardonnay, Australia

Sabbatical wines relate to places a good vintage has been consumed with friends. Our recent trip to Glen Lyon involved each of our party being responsible for one meal and suitable wines to accompany the food. A Vasse Felix Chardonnay being one of these – previously consumed in Australia with Aussie mates, hence highly pertinent.

WEEK 42
ALNWICK, WARKWORTH, & DUNSTANBURGH NORTHUMBRIA

It's 42 weeks since Granny and I set off on our 'Sabbatical' journey.

Possibly a good juncture to reiterate how this came about and what inspired us to embark on an unlikely project. In our letter to the grandchildren, we explained the timeline goes further back – now almost 8 years since my brother-in-law (David) died from a brain tumour and my cousins (Tim and Pete) both succumbed to pancreatic cancer.

This week, a reminder as to the importance and respect we should all attribute to good health: in the last week I've written 2 letters of condolence, had lunch with a good friend who has Parkinson's and the onset of dementia, plus a long-time friend and colleague called in and explained his inspiring and stoic approach to dealing with throat cancer.

If ever a kick up the backside was needed as to the importance of good health and counting one's blessings, this was the perfect reminder. The attitude and example of those with major issues generally puts one's own in perspective.

Hence Granny and my decision to recount a trip down memory lane, embarked on at the latter part of the summer involving 3 Northumbrian Castles: Alnwick, Warkworth, and Dunstanburgh. We're lucky enough to live in a beautiful part of the country, and the pandemic has made us look ever more closely as to what's in our own backyard.

The 1st image demonstrates the splendour, scale and magnificence of Alnwick Castle; sadly part of this obscured by scaffolding. Over the years we've visited the Castle on numerous occasions and for a variety of events. A memorable one being our eldest (your mum or aunt), a member of her school choir entertaining Princess Anne.

Our week 26 entry (Chillingham and Cragside) recounts a momentous charity bash at Chillingham Castle. The same committee held a further event a few years later at Alnwick Castle – another great night. Another outing which sticks in our memory relates to a Jools Holland concert, held in the grounds on the opposite bank of the river to the Castle. Another exceptional and never-to-be-forgotten night: the weather, music and company – brilliant.

The Jools Holland concert is embedded on our brain for another reason. Having returned home and in a deep sleep, I was woken by the phone and our son telling me "Dad! They're dead! The girls are dead!". I've never sobered up more in an instant - ever. A driver, under the influence of alcohol and drugs, had mown 2 of our son's friends down and catapulted them over his car. Luckily, by some miracle, both of them survived without suffering long-term injuries.

I tell the tale as it relates to this image, being the bridge over the river Aln. In Granny and my younger days, this was the main A1 – no bypass in those days. At sixteen I regularly travelled on my motorbike, to college at Ashington. I crossed the bridge on many occasions – although I seem to remember there being 2 lions, now only 1?

My story relates to my poor parents receiving a phone call in the middle of the night, and Mum hearing Dad agreeing to his son having a general anaesthetic. I'd managed to crash my motorbike and break my arm in 4 places, as well as incurring various lacerations. I tell the tale as a parent and grandparent – a salutary lesson for you grandchildren?

The next stage of our day involved travelling a dozen or so miles to Warkworth. Another magnificent castle, as can be seen from the image. The town of Warkworth is dominated by the castle, a beautiful and fascinating place. Great friends have had a static caravan outside Warkworth, overlooking the beach and North Sea for many years. We've visited on numerous occasions, hence frequented a number of Warkworth's restaurants and hostelries.

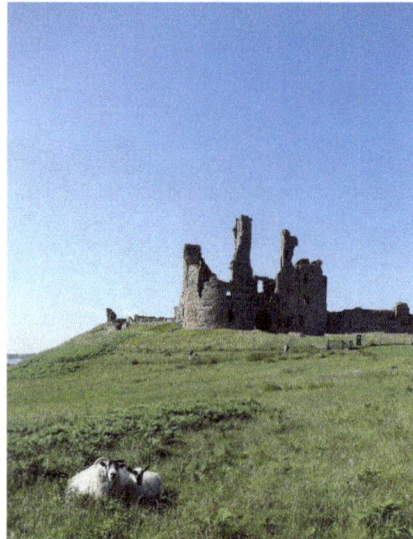

Our final destination Dunstanburgh Castle, a short distance up the coast beyond Craster. The 3 Castles demonstrate one which is inhabited (Alnwick), the next involves a ruin but incorporates aspects of how it once functioned (Warkworth), the third magnificent remains on a truly exceptional site overlooking the sea (Dunstanburgh).

A visit to Dunstanburgh Castle involves walking along the seashore from Craster, as demonstrated by the final image. There cannot be many more dramatic rambles, especially on a day like the one of our visit, the weather playing ball.

The depth, breadth and wealth of history in Northumbria well worth examination.

Authors & Books
With The End In Mind – Kathryn Mannix

During this last week the House of Lords have been debating a bill relating to Assisted Suicide. Granny and I have long held beliefs regards self-determination and individual choice. We are believers in the right to a Dignified Death. Kathryn Mannix may not, but her inspired book reference Hospice and Palliative Care in the NE an insightful read. When in Alnwick visit a spectacular book emporium 'Barter Books'.

Musicians & Popular Music
Lene Marlin – A Place Nearby

Lene Marlin; a singer introduced to us by our daughter. A powerful, highly emotional and evocative song invoking memories of friends no longer with us, plus an insight to aspects encountered on their journey and those close.

Composers & Classical Music
Phamie Gow – War Song

Phamie Gow's War Song fits this week's entry not only in relation to the wars and battles that have taken place over centuries involving all 3 Castles. But also the composition embraces aspects of the philosophical content.

Fine Wine
2003 Château Gruaud Larose, 2eme Cru Classe, Saint-Julien, Bordeaux, France

Family and relationships are a major feature and element of the Sabbatical message. Hence this week's wine choice remembers a dinner at home in the Borders with family and friends in 2014. Nothing better than good food, good wine and good company ~ Granny and I subscribe to that.

179

Granny

and I believe if ever a TV programme should be made compulsory viewing for all UK Citizens, it would be this: The annual Pride of Britain Awards. It humbles and inspires in equal quantities.

Watching the Awards last night we were blown away by: -

The inspirational and resilient young girl with legs and hands amputated due to meningitis, the young friends - one going through cancer, the other running to raise money - the joy of the Downs Syndrome family and their brilliant campaign, the mum whose son was hacked to death and her fight to help others in his name, the fireman who saved a drowning child submerged for 20 minutes, the triple amputee military veteran inspiring others, the boy who's camped out since the onset of COVID-19 generating vital funds for his local hospice, the young Syrian refugee assisting others via the Princes Trust, the dad running a football league to encourage men to talk about the loss of a child, the mum whose son succumbed to a brain tumour and driving force behind the UK's Organ Donor programme.

We appear to be living in an ever more polarised World – the 'Pride of Britain Awards' make one stop and think.

A further significant event took place this week in Glasgow: COP26 – World leaders gathering in an attempt to find solutions to climate change. Our 95 year old monarch (Queen Elizabeth) made an insightful statement: *"What leaders do for their people today is government and politics. What they do for the people of tomorrow; that is statesmanship"*.

It's interesting to look at our Prime Minister (Boris Johnston) and the President of France (Emmanuel Macron) and their current rhetoric – politics or statesmanship? Whipping up antagonism between Brits and our French neighbours, something Granny and I find distasteful. We're very fond of France and the French.

Our trips to Provence began before we were married, and have continued intermittently right up to the pandemic. This first image relates to our most recent visit and a birthday present – a flight in a glider over the Var region of Provence. Granny's sister (your great aunt) and late husband purchased a villa in Fayence many years ago, and we've been lucky enough to visit on numerous occasions. Patsy and Eddy's (Ab Fab) road trips through France legendary.

This my second trip in a glider from Fayence. The first memorable in that my pilot was a fascinating character; a vet whose various marriages had failed, hence no funds and living in a caravan. He'd worked around the World as far afield and remote as Alaska, the gliding club enabling him to continue his passion for flying. We'd such a great time, Jacques demonstrating the glider's capabilities, we far exceeded our allotted flight time - this rather perturbed Granny.

This picture looks inland from Fayence and brings back many happy memories for us both. During our early days, money was in extremely short supply. My cousin Pete (mentioned in our letter to you Grandchildren) very kindly lent us his 'Shack in the Woods' near Entrecasteaux. We'd a number of years with an annual trip to Provence. The final one when Granny was pregnant with our first child (your Mum or Aunt) – the facilities unsuitable for babies.

Almost 40 years since we last visited La Bergerie and no digital imagery available; it didn't exist at that time. Hence a picture supplied by Pete's wife on a recent visit and shows the Chateau at Entrecasteaux. We'd some wild and memorable escapades with an eclectic group of French friends whom we got to know over the years. The village would appear to have changed very little, and somewhere Granny and I must revisit at some stage.

Space as ever limited, but I should point out our annual Provence trip involved driving from the English/Scottish Borders, a ferry across the Channel and onwards to the South of France – quite a journey, and the road network nothing like today. Memorable stopovers in the likes of Paris, Marseille, Geneva and others.

At that time, days out to the Cote D'Azur from Entracasteaux involved travelling to St Maxime and Saint Tropez – the traffic horrendous. Hence we often drove inland to the Lac St Croix and Gorge du Verdon. Once again, no pictures from that time, but by pure fluke Granny's sister went 2 weeks ago and provided these memorable images.

Brings back memories of times gone by – my Subaru pick-up proving a most useful mode of transport.

Authors & Books

An Officer and A Spy – Robert Harris

Robert Harris' novel based on the true story of Georges Picquart a French military officer and his struggle to expose the truth of trumped up evidence which sent Alfred Drefus to the infamous Devil's Island.

Musicians & Popular Music

Edith Piaf – Non Je Ne Regret Rien

The title says it all – Granny and I have no regrets and count our many blessings.

Composers & Classical Music

Ennio Morricone – Chi Mai

Chi Mai takes me back 45+ years and the Lac St Croix. It conjures up a picture of lying on the lake shore, looking up at the Gorge du Verdon and hang gliders launching themselves off the side of the mountain, cavorting in the sky.

Fine Wine

2014 Olivier Leflaive, Auxey-Duresses, Cote de Beaune, Burgundy, France

Patsy and Eddy's road trips inevitably involve a night at Olivier Leflaive's hotel and a visit to his vineyard. Both partial to a nice French Burgundy, hence a wine consumed on one of their visits, but also with family in the Lakes in 2017.

183

WEEK 44
TINTAGEL, LAUNCELLS, & MORETONHAMPSTEAD ENGLAND

What

links King Arthur, Tintagel, Cornwall and attending our nephew's wedding 3 weeks ago? This taking place 44 years to the week Granny and I were married in Newcastle-upon-Tyne, honeymooning in Cornwall and Devon.

Extremely apt, this week's Sabbatical entry - 44.

Our first image features King Arthur, to whom we will return. Prior to visiting him, we drove from the Borders to Cornwall – a long way. On our journey we received a telephone call laying the foundations that our accommodation may be a tad suspect – described as 'Faded Grandeur'. Also a suggestion we should divert off the motorway to purchase towels, as there was a dearth of such. Our stay now logged in the annals of family folklore.

The 'Old Rectory'; an impressive building with loads of potential, but lacking more recent investment. Two dozen staying, our immediate family in the 'Coach House' next door. On arrival, we found the WC in the upstairs bathroom barred from use, wrapped in black insulating tape. One WC and a further downstairs shower-room WC for 10 guests laid the foundations as to the state of the facilities. We mucked in and got on with it – a great time had by all.

Launcells, a beautiful Church in the Cornish countryside, hosted a very special wedding. Two young people just made for each other and evident to everyone in the congregation. Granny and I'd only met the bride once, 3 years earlier at his brother's wedding (another special event). I remember telling our nephew – she's the one for you. Watching the facial expression and body language of both during the ceremony, incredibly powerful and an absolute delight.

Took Granny and I back 44 years and to the commitment we made to one another. This image of the newly-weds demonstrates the joy of the occasion. Epitomises how Granny and I felt and still do - we wouldn't change a thing. An aspect of the service deserving mention, my brother's (your great uncle's) beautiful solo rendition of Ave Maria.

The reception took place in a wedding venue a short walk from the church. The weather playing ball, enabling us to partake of drinks in the courtyard, with a lone balladeer serenading us. A special element of the wedding related to the guests of both bride and groom interacting with one another and new friendships forged.

After copious quantities of champagne, we sat down for the reception, meal and speeches. Excellent heartfelt speeches from the bride's dad, the groom, groom's younger brother (the best man), and the piece de la resistance, the bride. The information imparted left us in no doubt these two where made for one another. The band proved equally memorable, and everyone up on their feet boogieing the night away.

The following day, a lunch for 50 at the 'Old Rectory' - in the barn promoted as ideal for such. It had been suggested the owners might move the assorted eclectic paraphernalia inhabiting the barn, and run a broom over the floor. Family and friends at the 'Old Rectory' took it upon themselves to decant the tables and chairs from the Barn into the garden, the hog roast and bright autumn sunshine made for a further very special and memorable event.

A great day brought to a close with bride and groom leaving in their self-converted VW van – Denise 'Van' Outen.

The following morning we left and I thought a trip down memory lane in order. Tintagel, King Arthur's fabled castle only 30 miles away and somewhere we'd visited on our honeymoon 44 years earlier. I was also intrigued to see the new award-winning suspension bridge built since our original visit, which fulfilled my expectations.

My romantic magical mystery tour hampered by torrential rain and a howling gale. Almost blown off our feet and drenched to the skin. A Cornish pasty and café latte with steam rising from our wet attire did little to raise Granny's spirits. Especially as she'd

absolutely no recollection whatsoever of our prior visit to Tintagel!

The second surprise on my itinerary an equal failure. We spent the first 4 nights of our honeymoon at a hotel near Moretonhampstead. I thought a return visit in order. Unfortunately, the quaint Devon village of our memories not so in the wind and rain - hence no imagery. Eventually redeemed myself after establishing our hotel still in existence under a new name: Bovey Castle. Now very upmarket, an extortionately-priced sandwich and coffee doing the trick.

Authors & Books
Billy – Pamela Stephenson

I like people who inspire and make me laugh; Billy Connolly my all-time favourite comedian does both. Not sure his career could follow the same format in today's World – a sad predicament. From humble beginnings in Glasgow to a global stage. His wife's insightful telling of his story worth reading.

Musicians & Popular Music
Lindisfarne – Meet Me on the Corner

Lindisfarne, a Geordie folk rock band dating back to the late 60s/early 70s – very much part of Granny and my era and formative years. The opening lyrics 'Hey Mr Dream Seller' a powerful insight to our long and fruitful relationship.

Composers & Classical Music
Jeremiah Clarke – The Prince of Denmark's March

One of my dinner companions at the reception Gillie, a long-time friend of the Bride's family and organist at the Church ceremony earlier in the day; we covered a wide and diverse range of topics, the music to which the bride entered the church being one: The Prince of Denmark's March. Invokes memories of my Dad (your great grandpa).

Fine Wine
2019 Domaine Thomas-Collardot, Bourgogne Blanc Les Petits Poiriers, Burgundy, France

My brother asked me to supply wine for the select few staying for dinner on the last night. He and my sister-in-law abstemious by this stage – but having found the half-bottle left by my nephew (their youngest) and I, consumed it the following day. Praise indeed, the comment coming back: 'Paying that bit more noticeably worth it'.

BARRAGUNDA &
KENNETT RIVER
AUSTRALIA

This last week Granny and I should have been in Australia to celebrate a 70th birthday in Byron Bay. Unfortunately COVID-19 restrictions meant participation only possible via Zoom. An extremely poor substitute given their blue sky, blue sea, golden sands, great surf and temperatures in total contrast to ours. Never mind a party taking place.

The importance of family and friends an aspect of 'The Sabbatical' we wish to impart to our grandchildren. Friendship with the birthday girl goes back almost 50 years, when she married my cousin David. Sadly he was killed in a tragic accident a long time ago. Our friendships continued and we've embarked on many memorable escapades.

A perfect example being this week's Sabbatical entry. Our trip began in Melbourne and we travelled north to Cape Schanck to stay at Barragunda; an imposing stone mansion owned by a good friend. Barragunda has a fascinating history and equivalent to a Grade 1 listed building in the UK. Something I subsequently learned: the surnames of both my parents (your great grandparents) appear in the history and story of Barragunda.

Four of us had the run of the house and spread for a couple of days – memorable for all sorts of reasons. Even had the access code to the wine cellar, on one condition: we didn't drink the Penfolds Grange. Our host and hostess having previously joined us on the Laxford (Scotland) - even laid on fly fishing. But the highlight for Granny, her introduction to quad bikes – extremely nervous at the beginning, but couldn't be stopped once underway.

As can be deduced from our first image: Cape Schanck is stunning, and we had the place to ourselves! I attempted to get an image of a large gathering of kangaroos – but even on quad bikes they could far outrun us, their speed quite something to behold. The second image demonstrates the dramatic coast line and beaches – spectacular.

This third image highlights the facilities were far from shabby – the beautiful swimming pool available for our exclusive use. Before moving on to the second leg of our trip, we should record an extremely important night at the nearby Portsea Pub – a local institution and landmark. Mine host at the time, Andrew, a number of years later our companion and pilot when flying around the Outback in his aeroplane (see Sabbatical entries for Weeks 2, 8 and 12).

The second leg of our journey involved a ferry crossing from the Mornington Peninsular, as we were heading down to Kennet River on the Great Ocean Road – saved a lot of miles. An interesting garden ornament caught our eye on the journey. Someone crashed their private plane and had the tail-end built into the side of a hill - in their garden.

The house at Kennett River where we were staying; originally the holiday shack of my Cousin David's great university friend (the Doc) and his family. Sadly a further tragedy, in that he lost his wife to cancer. The amazing addendum to 2 such sad stories: the Doc and my cousin David's widow married, and we've had many amazing adventures together.

They rebuilt the shack at Kennett into an impressive holiday home, which the children could utilise. A very special site at the foothills of the Ottaway Rainforest, looking out to sea. We've been lucky enough to enjoy the house, and the many facilities on its doorstep, on a couple of occasions - boogie boarding one such pastime.

The Great Ocean Road is World-famous and wends its way along the shoreline, just beyond our picture of the unusual rock formations to be found at the beach. Not much further along the road are the Twelve Apostles, a reknowned spectacle – an interesting snippet learned about this on one of our visits.

A couple were playing away from home and had walked out over the rock archways. Unfortunately one of these collapsed, and they on the wrong side. Much explaining to be done when they appeared on the National News!

The final 2 images provide an insight to the varied wild life on the doorstep. Who needs David Attenborough when you can sit on the balcony with a 'tinny' and receive an education relating to Koalas and Kookaburras?

One enterprising Kookaburra provided a lesson in how they kill their prey prior to consumption. The point being, it was bacon and long dead, but still important to go through the motions. We also learned Koalas are known for 3 things: eating, sleeping and sex. Hence chlamydia a major issue. Amazing what one learns on your travels.

Authors & Books
Red Notice – Bill Browder

Political leaders around the World appear to be flexing their muscles and potentially playing with fire. None more so than Vladimir Putin of Russia. Red Notice provides an insight to the abuse of power and corruption in Russia.

Musicians & Popular Music
Freya Riding – Lost Without You

On Friday night, Granny and I watched the annual 'Children In Need' TV appeal. A young lad sang 'Lost Without You' in memory of his Dad - whom he'd lost to cancer. Also a tribute to the charity who'd encouraged his participation in music. Granny announcing Freya Riding's song one of her all-time favourites.

Composers & Classical Music
Chopin – Nocturne No 2 in E Minor

A thought-provoking piece of music which invokes contemplation; something 'The Sabbatical' aims to promote. Looking back on great times and memories, as well as counting one's blessings at the time of a world pandemic - helps one put things in perspective. Barragunda and Kennett River perfect ingredients to achieving such.

Fine Wine
2016 Penfolds Grange, Bin 95, Australia

Penfolds Grange, the forbidden fruit on our visit to Barragunda. Hence now in the Cellar Plan awaiting a visit from various Aussie Mates to the UK - so we can find out what we were missing. Hopefully comes to fruition in 2022?

Two

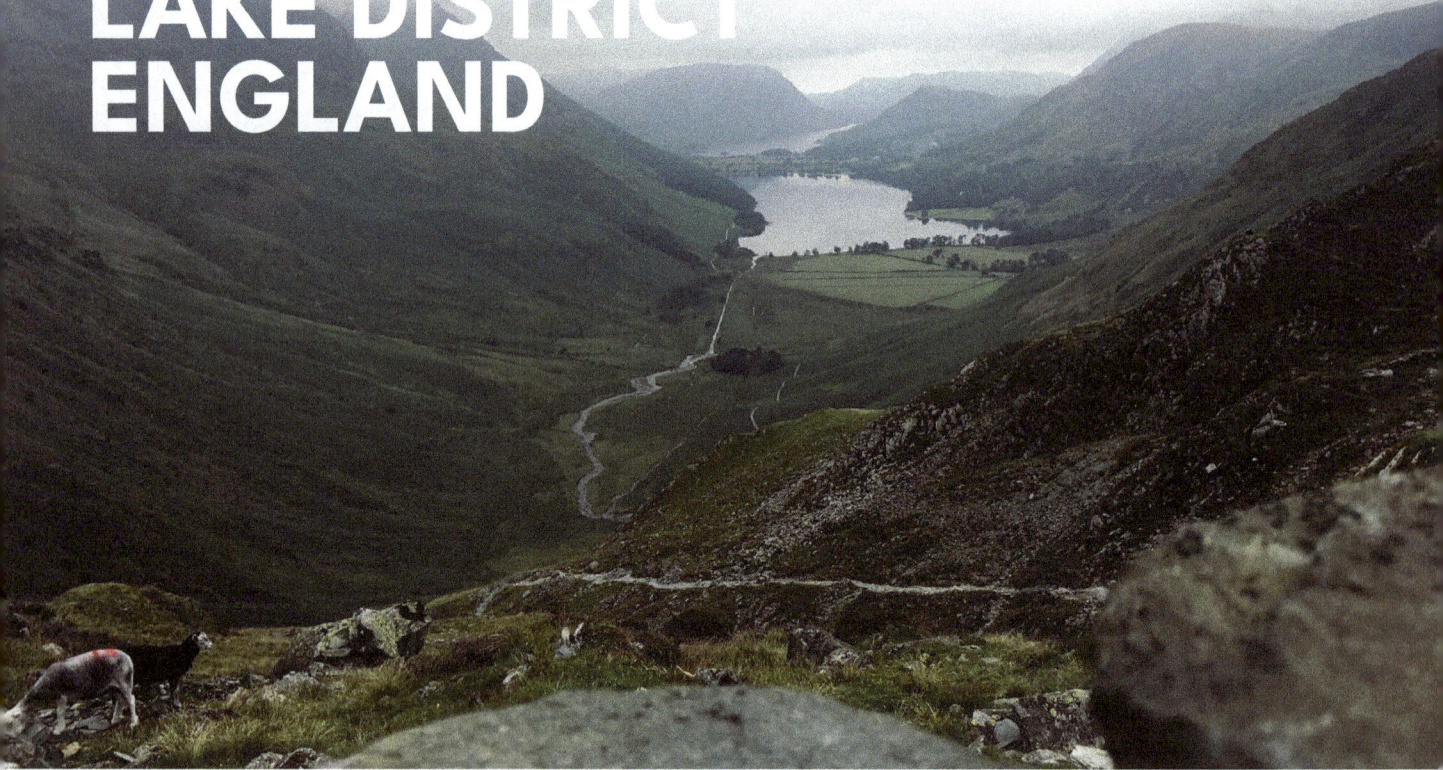

Two weeks back, Granny and I were in the Lake District to mark a special and sombre moment. The 11th day of the 11th month now has added significance for our family. Twelve months prior to the day, Granny lost her beloved Dad in difficult circumstances due to COVID-19. Donald truly a **gentle man** and a **gentleman** – **he** is sorely missed.

A good innings at 95, but the last few years of his life blighted by a stroke and vascular dementia. His final months involved a care home where he was well looked after - but sadly succumbed to COVID-19. Our trip to the Lakes involved a fitting ceremony and burying the second half of his ashes in the cemetery at Troutbeck. The exact same churchyard our brother-in-law David (mentioned in our original letter to you grandchildren) is also interred.

Donald and David had a great affinity for the Lakes. Donald's dating back to the 2nd World War when his school were evacuated from Newcastle to Windermere to escape the Nazi bombing of Tyneside. Donald billeted with the Vicar of Windermere – he'd fond memories of this time, and developed a life-long passion for the Lakes and mountains.

David and your great aunt's love of the Lakes grew via business – taking on a rather splendid Arts and Crafts B&B, originally Lord Lonsdale's shooting lodge, and over a period of 20+ years developing this into a multi-award winning Michelin-starred hotel. Which, as you will hear, played a large part in our love of the Lakes and memorable times.

Many momentous family occasions have taken place in the Lakes - and over many years. Celebrations involving great joy and sadness. Our recent visit to Troutbeck brought back memories of David's funeral, held at the same beautiful Lakeland Church. His memorial took place in the much larger venue of St Mary's Windermere. Lasting memories: 2 of his great mates doing a double act (very Morecombe and Wise), plus the amazing opera singer – special tributes.

A stand-out moment relates to our No.1 Son's engagement. I collected them from near Lake Windermere and came back via Troutbeck for champagne at your great aunt's – we've a lovely image of the happy couple near here. The same vantage point

at another time, we watched the 2 surviving airworthy Lancaster bombers from the 2nd World War fly along Lake Windermere, in celebration of the daring Dam Busters raid on Germany – a spectacular sight.

I feel I should explain the initial images: the first a typical Lakes vista looking down towards Buttermere - as can be deduced, the Lake District is spectacular and awe-inspiring. The second just off the road between Lake Windermere and Troutbeck – incorporates a stereotypical view of the Lakes - dry stone dykes, pastures, sheep and inevitably a lake.

This next image is more personal, and relates to the many visits paid to the Low-Wood leisure complex overlooking Lake Windermere with our 2 youngsters. Our No.1 Daughter loved nothing better than swimming, followed by a dip in the ice-cold plunge-pool reserved for those who'd had a sauna or Turkish bath –not sure her brother was quite as keen. The spa and a relaxing massage at a time when life was highly pressured was enjoyed by Granny and I.

An abiding memory at a slightly earlier time: Granny and I enjoying an up-market dinner while staying at our in-laws' establishment. Two forlorn little figures being brought to the dining room, holding hands along with a teddy bear – we being asked if they belonged to us, our youngsters having been found wandering the corridors.

We frequented the hotel on many occasions, rounding up a group of mates to support the family venture. I believe this was more welcome in the early days, as they went up-market and more successful, possibly seen as a touch too lively; hence always given the last slot at dinner, and our section of the restaurant closed off from other guests.

An aspect of our family's involvement with the hotel relates to our son (your Dad or uncle) deciding to drop out of his university course – he'd the good sense to resume the following year studying Modern History and Politics. This involved an extended period prior to returning to Loughborough. Granny and I decided he'd have to earn a living until then. Hence he ended up working at the hotel, this the introduction to his future career – fine wine.

Many family celebrations and gatherings took place at the hotel over the years: too many to mention, but all memorable. The final few images hopefully provide an indication as to how special the Lakes are to us and our family. Granny's sister (your great aunt) resides in the Lakes – hence we still enjoy visiting on a regular basis.

Authors & Books
A Shepherd's Life – James Rebanks

A fascinating insight into farming life in the Lake District – but also an inspiring read as to someone who struggled with his early education, but transcended this and later attended Oxford University. Much to be learned from this book.

Musicians & Popular Music
ELO – Mr Blue Skies

Only one possible song and band for this week's Sabbatical entry: the Electric Light Orchestra's Mr Blue Skies - played at our brother-in-law David's (your great uncle's) memorial service, and one of his great favourites.

Composers & Classical Music
Gabriel Faure – Requiem: In Paradisum

Chosen by Granny for a variety of reasons: as a member of our local church choir she performed this beautiful choral piece, a great favourite of hers. Extremely apt in relation to her Dad (Donald) and brother-in-law (David).

Fine Wine
NV Billecart-Salmon Brut Rosé, Champagne, France

Relates to 2009, and a special wine tasting in the Lake District which involved a number of great mates who have featured in the Sabbatical. The Billecart-Salmon provided by our brother-in-law David. A favourite of his, and excellent.

Granny

and my experience of the Dalaman Coast in Turkey stretches back 35 years. At the time, life hectic from both a family and work perspective. Two young children and a business in its infancy. We'd dinner with a good friend, a travel agent who'd just returned from sailing on a Gullet in Turkey, and came highly recommended.

Drink, bravado and stating 'that sounds great' – all the incentive Granny required to book, and 2 weeks later we were in Marmaris. It's worth pointing out, in those days you flew into an inland military airfield – a hairy landing and equally scary bus ride. On the dockside we met Mammet and crew mates – proved delightful. We the first to arrive.

No concept of our fellow travellers, so when the first joined us: a rather short, stocky gentleman, carrying an excess of weight, dressed in a bright red tracksuit, a broad cockney accent and hailing from 'Soff-End' (Southend). We weren't sure what we'd let ourselves in for. Over the week he and his wife proved charming, and the stewards at Southend Conservative Club. He was an interesting character, having been both a professional wrestler and masseuse.

The point being, never judge a book by its cover or a fellow human being on first appearance. We and our 8 fellow travellers proved the most eclectic bunch of souls you've ever met and got on like a house on-fire, and had a blast.

Lots of shenanigans and hijinks proving the order of the day.

A memorable excursion involved transferring to a smaller craft and travelling up a river. We saw the most amazing Roman architecture, including an almost intact amphitheatre - completely unattended and deserted. After a memorable banquet in a local restaurant and copious quantities of wine, we ended up in the hot springs on the shore of an enormous inland lake (once again totally deserted) – the smell of sulphur lodged in the memory.

Too many tales could be told, but one should be on the agenda: fun, hilarity and extracting the Michael took place from start to finish of this memorable trip. Revenge taken on one of our number (who will remain nameless). Obviously taken things to an extreme, so ended up strapped naked to the mast. One's actions can involve consequences.

Another lesson in life – always treat fellow humans with courtesy and respect. The previous guests on the boat were German and aboard for 2 weeks. Turned out they'd recently seen the film 'Death on the Orient Express' which made then think all Turks were rapists, murderers and drug dealers. They refused to acknowledge Mammet and team – behaviour abysmal and rude. After a week got the washing-up liquid in the meatballs - confined to their bunks.

The lesson being, always treat your fellow human with good grace and manners. No such ramifications from our food - excellent.

Our second trip to Turkey, and once more on a Gullet, took place a decade later. This voyage included our 2 youngsters – the eldest in early teens at that time. We sailed on a beautifully-built triple-masted Gullet, another memorable holiday. Our No.1 Daughter and No.1 Son were/are exceptional swimmers, and we spent much time in the water – seeing and following giant turtles a major highlight. Once again, we knew none of our fellow guests; this proved entertaining and fun.

A story worth recalling related to a lovely older couple. I came on deck to find she was waving to her husband, who was frantically waving back - about 100 or so metres from the boat. We realised he was in trouble, having encountered serious cramp. Our daughter was with me, so we dived in, and those school life-saving lessons proved their worth.

Our third trip to Turkey took place the summer prior to the pandemic – also on a boat; this time a mate's gin-palace. We've no pictures of our earlier trips to Turkey – before digital photography. So all pictures from this last trip or courtesy of the next. Once again a memorable voyage involving members of the 'A' Team – refer back to Week 37.

As can be deduced from the imagery, the Turkish coast and scenery is stupendous, and the Turkish people we've encountered amazing. As ever, not enough space to record all our many exploits – save to say we had a ball. So much so; a special trip down memory lane planned for next May – 10 of us on a Gullet: the 'A' Team on Tour in Turkey.

The final image shows 'Salamander', which will be our home for the week – Granny and I can't wait.

Authors & Books
Memoirs of a Geisha – Arthur Golden

A book I read on our last voyage. Given the focus on modern day slavery at this time: from impoverished fisherman's daughter to final years in the Astoria New York – a fascinating insight to a different time and culture.

Musicians & Popular Music
Nina Simone – Feeling Good

Granny appears to be taking control of the popular music choices – this from her playlist. The Sabbatical aims to be positive, and hopefully insightful – Feeling Good and its lyrics extremely apt, and also a favourite of mine.

Composers & Classical Music
Gustav Holst – Mars from the Planets

There's no better place to observe the stars and planets than when at anchor in a secluded bay and no light pollution. Holst's Planets Suite epitomises some very special and memorable voyages we've enjoyed over the years.

Fine Wine
2020 Lismore Estate Vineyards, Chenin Blanc, Cape South Coast, South Africa

No record of any of the wines consumed in Turkey – possibly the volume an influence. So have chosen a new wine Granny and I were introduced to this last week. Granny often turns her nose up at a Chenin Blanc – not this one.

WEEK 48
EDINBURGH SCOTLAND

A week ago on Friday, Granny and I attended a dinner in Edinburgh – A 'Retirement Celebration' in honour of me. Having worked alongside an Edinburgh company founded from scratch some 27 years ago, we've developed a special relationship and partnership which they wanted to acknowledge. I was blown away and humbled by this.

At the dinner I said a few words which give an insight to my involvement with Edinburgh – going back many years:

Part of my education took place less than a mile from here, at one of Scotland's prestigious Public Schools – almost 3 years from the age of 13. The less said about this the better, as my time at the institution came to a premature end.

One of the few positive aspects of my time in education was just up the road from here, at Basil Patterson's: an educational establishment for rejects. I spent 2 terms cramming for 'A' level resits. A remarkable lady taught me – Mrs McDowell. My first tutorial involved précising a piece and being duly informed I'd no possible chance of attaining an 'A' level – 'But she would teach me to appreciate English, marshal my thoughts and to put these down on paper'.

Elements of what you, our grandchildren, have just learned are hopefully what not to do. But it paints a picture as to my early involvement with Edinburgh, and subsequently Granny's.

During my 2 terms at Basil Patterson's, I lodged with my aunt and uncle in Edinburgh. She was a lovely lady with a quiet, mild manner, but a well-developed intellect. My uncle (my Mum's youngest brother) worked at the time in Borneo and Sarawak – involved in developing the Kula National Park. We got to know one another, and the second image features the Botanical Gardens, where he introduced me to plants and trees he'd brought back from his travels.

My uncle (Your great, great uncle) was a quiet, deep thinker, which I believe related to his involvement as a young, highly-decorated officer in the Black Watch during WW2 –a very brave man. During my late teens and early 20s, whenever he visited, we'd invariably go for a pint. Sadly 'Weils Disease', caught while in the Far East, was to blight his life; he was lucky to survive, but suffered major health ramifications.

201

In more recent times, Edinburgh has played a large role in our lives – our eldest and family residing in the city. The Botanical Gardens a short walk from their home – many enjoyable times chasing our first grandchild (JJ) around the gardens. Since moved to the outskirts, produced our first Granddaughter (Loz) – and many more fun times together.

This image takes me back to my childhood and visits to Edinburgh's Military Tattoo – invariably in pouring rain (the venue just visible in the top right-hand corner). My Mum and Dad (your great grandparents), as I've previously espoused, had active involvement in the military during WW2 and saw the Tattoo as a celebration of lost comrades.

Memorable meals appear to be an aspect of our involvement with Edinburgh - this tale also relates to the Castle. It's not every day you're entertained in the Officer's Mess at Edinburgh Castle by a Brigadier General. The Sabbatical entry for Week 30 (Glen Lyon and Loch Tay) is how this came about. On a visit to Glen Lyon we were invited to partake in a ceilidh raising funds for Cancer Research – a lovely lady from Fortingall undergoing treatment at the time.

Her husband was a fellow officer of Brigadier General Andrew MacKay. We won a raffle prize to have lunch with Andrew at Edinburgh Castle. An interesting aspect of our day related to reporting to the military entrance. The sergeant in charge took one look and assumed we were misplaced tourists. "And what can I do for you, sir?" he asked with mock sincerity. On stating we were there to see the Brigadier General – a sharp click of the heels, a salute, and "Yes, Sir!"

This picture of Chamber Street Museum takes me back to my childhood; at around 10 years of age, my oldest friend in life and I would get on the train at Berwick for a day out in Edinburgh. This one of our favourite haunts.

Many more tales of Edinburgh could be told, but simply not enough space. A memorable 40th birthday bash at Murrayfield: Celine Dion and The Corrs. Countless Rugby Internationals at the same venue. My son-in-law converting the winning try for Berwick. Opening nights of the Impresario's musicals: Crusade, Eurobeat and Connect. A special wedding at the Signet Library. Last night of the Rocky Horror Show – too much to recount.

Authors & Books
The Edge of Eternity – Ken Follett

This book covers 2/3rds of Granny and my lives – and encompasses World events from prior to erecting the Berlin Wall to the inauguration of Barack Obama. A noticeable absentee at last Friday's dinner was Suzanne, who played an important role in the building of their business. She crossed the Wall when the consequence often led to death.

Musicians & Popular Music
James Taylor – Fire and Rain

For me, this song conjures up Suzanne, and included as a tribute to her. A formidable and tenacious lady with a razor-sharp mind and wit. A friend who sadly took her own life. The last person I'd ever have imagined doing so. The refreshing approach to mental health compared to Granny and my younger days is to be applauded.

Composers & Classical Music
Franz Liszt – Consolation No3

An evocative and moving composition. Fitting in relation to Edinburgh regards visits to Morton Hall over the years – both my parents' (your great grandparents) cremation services held at this location, as was Suzanne's and others.

Fine Wine
NV Laurent-Perrier, Grand Siècle No 24, Brut, Champagne, France

We've just received a special order of Champagne – some for Christmas, but also a bottle on ice as our fourth grandchild is due at any moment. Granny and I waiting with baited breath for news, as the due date has now been exceeded.

203

For this week's Sabbatical entry I had to trawl way back through our archives to find imagery from 20+ years ago. My first ever digital camera, and severely limited function compared to today's models. Granny berates me for my pictures being of scenery and not people. Not totally true, as will be demonstrated – pictures going back many years!

Great mates (the Don and Contessa) had a family apartment first at Dominion Beach, and later at Cabo Bermejo in San Pedro – many memorable holidays and magical mystery tours associated with these 2 locations over the years. It's hard to know where to start – so we decided the easiest way was via the 6 pictures we whittled our choice down to.

WEEK 49
SAN PEDRO, ESTEPONA, & RONDA, SPAIN

Turns out 1 or 2 of the images not suitable for publication from this particular excursion. Way up in the hills behind San Pedro: the secret river. Swimming the only access, so secluded and unknown to others at that time. Here a pool where we dived off the rocks, swam and had fun. A number of visits over the years to this special place.

Our host and hostess well-known for magical mystery tours, and this next image brings back memories of one of these. The picnic hamper prepared, plus the wine and beers in the cool box. We arrived at the marina in Estepona to be informed we were off on a boat, fishing. Unbeknown to us, they were taking delivery of their first-ever vessel. No inkling this would lead to many more memorable boat trips in a variety of countries over the next 20+ years.

Estepona became their home port for a series of 2 boats, and Granny and I lucky enough to accompany them as Cabin Boy and Dolly Deck Girl on a number of voyages over the years. The image of the dolphins taken on a memorable trip crossing one of the busiest shipping lanes in the World (the Straits of Gibraltar). We encountered a massive pod of dolphins on our voyage to Morocco, a magnificent sight – not easy to photograph from a small craft.

Too many boat trips to record, so will restrict this to one other. We recently wrote about our visit to Barragunda and Portsea (Week 45, Australia). Andrew, mine host at the Portsea Pub and later our companion and pilot around the outback, presented me with a much-prized Portsea Pub cap. Lost overboard on a boat trip to Gibraltar while encountering high seas and a strong wind – a beer much-welcomed by Granny when we made land.

An image highlighting another magical mystery tour involved a taxi ride to a local train station and a meandering journey up through the hills – the destination: Ronda. I'm presuming we'd run out of money as it would appear our wives were peddling their wares. Proof to Granny I take the odd picture including people – but the subject has to be worthy. Hopefully a demonstration we've encountered much fun and hijinks over the years.

This next image of Gaucin, a beautiful village up in the Mountains of Andalusia - one of our stopovers. This a memorable road trip encompassing Tarfia, Cadiz and others. Tarifa etched in my memory: we stopped for a beer at a beach-side bar (Hurricane's) - the girls visiting their shop, the Don and I left to watch the kite-surfers.

Turned out a fashion shoot taking place on the sands immediately in front of us. The models changing their designer outfits and showering on the beach adjacent to us – we, being the gentlemen we are, had to avert our gaze.

Alfonso, a local character, was encountered at one of our favourite restaurants in the hills above San Pedro. The Forge run by Mike and Athene; a fixture on our itinerary. Alfonso and his trusty donkey used to drop in for a beer. Eventually banned after passing out in the garden one night, the donkey wrecking their much-prized vegetable patch. Another visit involved a pogo-stick competition arranged by Mike and Athene's lovely granddaughter Ella – me being the winner.

The final image features the Alhambra in Granada – a magnificent series of palaces and fortifications dating back to the mid-14th Century. The complex is vast; fascinating to see the juxtaposition of Islamic/Moorish buildings mixed with later Catholic Spanish architecture. One of Spain's most popular tourist attractions – we much enjoyed our visit.

A final reminiscence of Cabo Bermejo relates to another favourite watering hole - Tikitano's. Many memorable Sunday lunches enjoyed with Frank playing the piano. The inevitable conclusion, 'Sex on the Beach' – a bright orange cocktail.

Authors & Books

Inside The Kingdom – Robert Lacey

We live in a time when it would appear people are quick to jump to conclusions. A visit to the Alhambra highlights the enlightened Islamic/Moorish society of that era. This book is worth reading - it provides a more recent insight.

Musicians & Popular Music

Madeleine Peyroux – Smile

Jazz hasn't featured prominently in the Sabbatical – but as highlighted in Week 5 (New York), we're fans. The song apt for 2 reasons: first, Madeleine Peyroux introduced to us at the Forge. Second, the song title is something we subscribe to.

Composers & Classical Music

Francisco Tarrega – Memories of the Alhambra (Classical Guitar)

Works in relation to the title – the Alhambra had a profound effect on Granny and me. Perfect choice regards classical guitar, and conjures up images of many memorable escapades and veritable feasts in Spain over a great many years.

Fine Wine

1985 Marques de Caceres, Gran Reserva, Rioja DOCa, Spain

I believe I've alluded to Granny being partial to a Spanish Rioja. Marques de Caceres a wine we've consumed both in Spain with friends, but also with family when supplied by our son (Christmas 2007). A gift from him to us after a busy first festive season working in the wine trade.

The Sabbatical fast approaching its conclusion – almost a year since Granny and I set off on our quest. The COVID-19 pandemic and lockdowns the original catalyst, with an uncomfortable feeling of déjà vu. Yesterday's daily infection rate at 91,743 and growing exponentially (+60.8% over the last 7 days). The plus side relates to deaths reported within 28 days of a positive test: 44 yesterday, the 7 day average: 5.4%, a fraction of those a year ago.

Rewind twelve months: Our Christmas celebration to be held in Edinburgh with our daughter, son-in-law and 2 eldest Grandchildren - cancelled when Sturgeon's government banned us from crossing the Border. The return match now postponed, as our eldest grandson has tested positive for COVID-19 and they have to isolate for 10 days.

On a much more positive note, we're celebrating the birth of our latest grandchild, born this last week. A boy, 1oz short of 10lbs – Mum and baby home and doing well. Puts everything in perspective; we are over the moon.

London played an important part in my development, and helped shape much of my attitude and approach to life. I was a wild child – rebellious, hot-headed, opinionated and with a strong belief in the importance of fun. In my late teens I dropped out of a business course at Newcastle Poly after 2 terms and headed to London. My academic career achieved very little, but the world of work propelled me into a new environment within which I thrived.

I became a Management Trainee for an exciting new venture – Europe's first hi-fi department store on Tottenham Court Road, London in the early 70s, and living with 12 other Management Trainees in the Maree Hotel, Gower Street - their staff consisting of a vivacious group of Spanish au-pairs. A World away from my previous existence.

To emphasise the fact life wasn't all about hard work and living fast and free: the British Museum not far, and a place I sometimes strolled through during my lunch break. The glazed atrium in the library rotunda area didn't exist 50 years ago – a piece of architecture I much admire. London's many museums and the history, education and knowledge to be drawn from these is something we'd much like to point out to you, our grandchildren.

After the Maree Hotel, I and a group of colleagues rented a flat off the New Kings Road in Chelsea – a very trendy place in the early 70s. Many great watering-holes and eateries on our doorstep. A much-favoured destination was the Duke of Cumberland, Parsons Green. At that time your great aunt (my sister) lived in a flat overlooking the green.

This next image shows Chiswick House – this immediately across the road from my next abode: The Old Manse, a large rambling house which I shared with an eclectic group of individuals. Chiswick was later to play a part in Granny's and our lives, as my cousin Tim (mentioned in my original letter to you grandchildren) lived here and we often stayed. Subsequent to this, great Aussie mates had a house in Chiswick and many more memorable visits.

Our son and daughter-in-law rented their property while renovating their own house in Peckham. Granny and I have fond memories of the area. A much-favoured place being Kew Gardens, not far away and where we came across the artwork of Rory McEwan (Week 34). Richmond Park and walks along the river to the Dove were other favoured haunts.

It would be remiss not to mention the many musicals and shows Granny and I have enjoyed while in London. Too many to mention, but here some highlights: Jesus Christ Superstar, Joseph, Showboat, Miss Saigon... all seen many years ago. More recently, Les Miserables, War Horse, Mamma Mia, Billy Elliot and Eurobeat, to name just a few.

The Cutty Sark has always fascinated me, as it was built for Top Hat Willis who emanated from our part of the World. I'm led to believe there is some sort of family connection. A magnificent vessel, and should be on your itinerary.

As ever, so little space to record many years of living and visits, so a couple of extra highlights: a walk in St James' Park, afternoon tea at the Goring, wine-tasting at Berry Bros. and Rudd, dinner at Café Morena - all in the one day!

Granny and I would like to finish with a favourite destination and particular event. The Opera House, Covent Garden and a memorable wedding. My cousin Tim's wedding held in the chapel at Covent Garden, reception in the Opera House. Men in top hat and tails, ladies in their finery – and Japanese tourists mistaking us for Royalty.

Authors & Books
**Rogue Warriors –
Richard Macinko and John Weisman**

The title appealed to Granny for some reason – I can't think why! The content paints a picture of the exploits of an eclectic and diverse group of young men and demonstrates courage and bravery above and beyond the call of duty.

Musicians & Popular Music
Ralph McTell – Streets of London

My sister (your great aunt) and a friend ran a catering company in London. I once helped out at a Sex Pistols function. More food thrown than eaten – so in the early hours, we took what was left and knocked on cardboard boxes under Waterloo Bridge. A connection to Ralph McTell's insightful composition and the plight of others.

Composers & Classical Music
**Edward Elgar –
Cello Concerto (Yo Yo Ma)**

Apt to utilise an English composer – an arrangement that conjures up a wide range of emotions. London has played an important part in our lives; Elgar's Cello Concerto is perfect to encompass a wide range of diverse experiences.

Fine Wine
**2015 Domaine Drouhin,
Pinot Noir, Dundee Hills, USA**

A wine drunk with family and friends at one of Granny's favourite restaurants in Chiswick - La Trompette. Been lucky enough to frequent this establishment on numerous occasions with a broad range of mates.

ASSORTED OZ ODYSSEYS AUSTRALIA

This being the penultimate 'Sabbatical' entry, Granny and I decided to return to Australia and highlight a number of further memorable 'Odysseys' that have failed to be incorporated – as up to this point. Travel broadens one's horizons, and Oz has certainly expanded our knowledge and insight regards many things – hence some additions.

Farms and Homesteads have appeared in various 'Oz Odysseys' throughout the Sabbatical: this image being of 'Mungo' (Week 12), a magical and ethereal place. Our stay at 'Blinman' (Week 2), a further enchanting outback homestead. Our lunch at 'The Farm' on the outskirts of Byron Bay (Week 33), another memorable moment.

One farm and homestead that has not graced the pages of the Sabbatical, as yet, is 'Karawan' (Victoria). We've visited and assisted (hindered) around the farm on a couple of occasions. Sticks in our memories for a variety of reasons.

First time we came across a 'Redback', one of Australia's many lethal spiders. An off-road trip from Kennet River (Week 45) to the 'Farm' through the Ottaway Rainforest. Granny reacquainted with quad bikes and running amok. The Aberdeen Angus bulls on the rampage - us in-between. The search for young calves, as featured in this next image. Outings to the pub 'The Gelibrand' for the amber nectar (up to the 60s, this the local railway station).

Also, a significant family reunion involving 4 generations. I've previously mentioned my Aunt Joan (your great, great aunt), uncle and 3 cousins all emigrated to Australia as £10.00 Poms in the 60s. I'd not seen her in a number of decades, but over the intervening years we'd enjoyed intermittent in-depth worldly chats via phone. She resided in Colac, not far from the 'Farm', and hosted a memorable family gathering - a very special lady and event.

The death this week of Archbishop Desmond Tutu brings her to mind. Desmond Tutu and Nelson Mandela were, in our opinion, the archetypal religious and political leaders our World currently lacks. These 2 were iconic, exceptional men. My aunt and uncle were anti-apartheid activists long before the cause gained traction in the West. It's fitting to mention Joan at a time when the World has lost a great example of humanity, humility, courage, kindness, and fun.

While on the subject of my aunt: Granny and Joan had never previously met. Granny couldn't believe how similar in mannerisms, voice and approach to life my father and aunt were. They'd spent very little of their lives together - but just like 2 peas in a pod. Both inspiring people, and well worth emulating - greatly missed to this day.

Something they'd have found of interest would be the debate around 'Cancel Culture' and 'Woke'. They believed in the importance of an open and enquiring mind. Our trips to 'Oz' have opened up new horizons and awareness for Granny and me. These 2 images highlight a diversity of life never encountered on our home patch. To which you can add: kangaroos, wallabies, emus, koalas, crocodiles, mantas, stingers, spiders, snakes, toads, and many others.

We've also been lucky enough to encounter people from all races, creeds, and colours – from wide-ranging and diverse backgrounds. Travel, and opening one's mind to the experience of others, are something we'd love to pass on.

This next image features 'Hanging Rock', visited on our first foray to Australia. Hanging Rock is an impressive prehistoric volcanic outcrop which dominates the area – the views over the plains spectacular. Like many places we've been to in Australia, it appears to possess a spiritual aura. The indigenous people's ownership dating back some 26,000 years. A much more recent mystery surrounding the Rock adds further drama – Picnic at Hanging Rock?

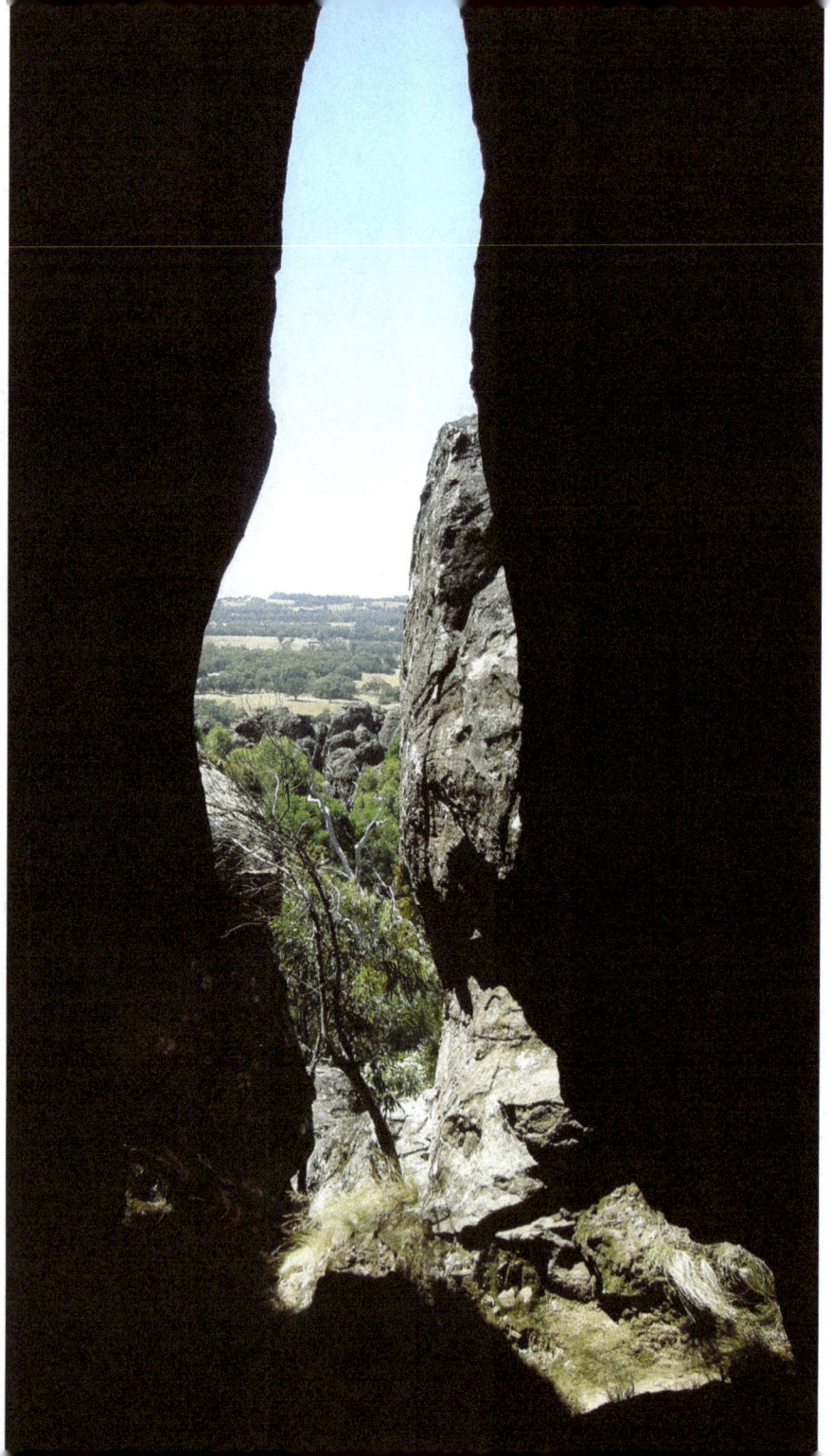

Granny and I can't finish without highlighting visits to 'Mooroolbark Mansion', the home of the Impresario mentioned within the Sabbatical on a number of occasions. These celebrity types get a little touchy about publicity; hence no photographic evidence. Suffice to say, looking out over the 'Dandenong Ranges' was a great experience – plus spectacular parties at the Mansion, as well as a memorable invitation to a nearby vineyard Domaine Chandon.

Our most recent visit to Oz culminated in a far-too-brief stopover in the Margaret River (Western Australia). As the image demonstrates, the weather was not playing ball. But our stay at Cape Lodge, visits to the Cape Mentelle and Leeuwin Estates wineries, a chauffeur-driven tour which included this spectacular beach. Hence Perth and the Margaret River very much on our itinerary for a further visit - whenever that becomes possible.

Authors & Books
Alec – Alec Wek

A short but powerful book: from Sudanese Refugee to international supermodel. Whenever you're feeling sorry for yourself, it's worth delving into the lives of others less fortunate. This lady is an example to us all, in so many ways.

Musicians & Popular Music
Super Tramp – The Logical Song

There are many bands and songs Granny and I could utilise, but as we're on the penultimate entry here's one that hopefully makes you (our grandchildren) stop and think. A long-time favourite - with thought-provoking lyrics.

Composers & Classical Music
Karl Jenkins – Benedictus

A composition that invokes memories of people and places – sadly not all still with us, but far from forgotten. I'm not religious, unlike Granny, but admire the faith of others (whatever that may be) and, more importantly, good deeds. Desmond Tutu a shining light and exemplification of these.

Fine Wine
2017 Benjamin Leroux Les Baudines, Chassagne-Montrachet Premier Cru, France

A fitting wine consumed in London (2019 and pre-pandemic) with 4 great Aussie mates who've been instrumental in many of our 'Oz Odysseys'. Even more apt, as we're planning a reunion in Cannes (France) in May – COVID-19 conditions prevailing.

WEEK 52
NORHAM & HORNCLIFFE ENGLISH/ SCOTTISH BORDERS

Fifty-Two

weeks, and this the final 'Sabbatical' entry! The original intro explained our aim and intention:

The 'Sabbatical' is dedicated to our grandchildren at a time of uncertainty and an attempt via imagery, places, words, people, memories, composers, books, musicians, and fine wine – to provide an antidote.

It's fascinating to look back on this time last year when in 'lockdown'. Twelve months on, we're in a state of limbo: the Omicron variant and number of people infected is through the roof, but hospitalisations and deaths are way below those of a year earlier. Will the answer prove political, scientific, or an amalgamation? The politicians at Westminster, as well as the devolved Nations, appear (to Granny and I) a tad too interested in politics than 'Statesmanship'.

In contrast, we've had the best New Year's present ever – having met our new Grandson for the first time. Added to which we've been with his elder brother (born early-on in the pandemic). Our other 2 grandchildren having tested positive meant Christmas was cancelled - but luckily no adverse health issues whatsoever. So all is well in our World.

Having reminisced on a wide array of holidays, places and memories from diverse parts of the globe throughout the past year, the question arose as to our final destination. Which proved easy - it's the place we love to return whenever we've been away. To assist in explaining more, we've split this into 2 halves: the first relates to my childhood in Horncliffe. The second, raising our 2 children (your Mum, Dad, aunt or uncle) after returning to the area.

Granny and I realise we're blessed to live in a special part of the UK – the English/Scottish Borders, particularly during a period of COVID-19 lockdowns and government restrictions. We're lucky enough to have a detached house, garden and beautiful countryside on our doorstep. We decided to highlight our lockdown walks via imagery. Norham to Horncliffe, or the other way round (both special) –and as you will see, they vary throughout the seasons.

How do you distil 7 decades of interaction with Norham and Horncliffe into a few hundred words? Impossible, but hopefully we can provide an insight: my Mum (your great granny) had a habit of inviting assorted waifs and strays, hence a lack of accommodation. After the 2nd World War, my uncle worked for the Forestry Commission and lived in a large Army surplus tent –which we inherited. Our summers spent camped out in the garden; no room at the inn.

John Archer, the local market gardener, later regaled us with stories of returning little people in their pyjamas, wandering around the village in the early hours, to oblivious adults. Great entertainment was climbing to the top of the tent and using it as a giant slide. I'd an interesting and exciting childhood, and no desire to alter anything.

As you can see, the river Tweed is on our doorstep and very much part of our playground. Teaching us to swim at an early age was a good idea, as we spent many hours in canoes and boats. Paddling 5 miles downriver to the sea without parental supervision not unusual. Would be frowned upon in today's society – but great fun at the time.

Running out of space, so will leave a couple of insights: my education commenced at Horncliffe School (long closed), built in 1832, and little had changed over the intervening years. The one thing we learned was to fight, demonstrated when my parents sent me at 7 to a school in Berwick for 2 terms before packing me off to boarding school.

My second week involved the school bullies cornering me – having removed one's front teeth, the other ran off. The headmaster shook me by the hand, and never another problem for the duration. On a different note, Mum and Dad held open house at New Year. One local lad, the worse for wear, passed out on the settee in the front hall. Mum placed a wreath on her Mum's grave every year - caused great consternation when laid on his chest!

We must break the 'Golden Rule' of not exceeding 1,000 words – this the final entry, and rules are for breaking!

Christmas Eve 39 years ago, we moved back to the district – our daughter about to celebrate her first birthday. Snow on the ground and well below freezing, and the heating in our new home failed. My brother-in-law gallantly came to the rescue and whisked my wife and daughter away. An inauspicious start to our family's life back in the Borders.

Squeezing 39 years into a meagre few paragraphs totally impossible - but once again, a few insights. Two years on and our son (your dad or uncle) came into the World. School a mere few miles from the house, and our 2 thrived – from 4 to 18. Showing up their parents by becoming Head Girl and Head Boy and gaining university degrees.

Sport played a massive part in their school days: swimming, hockey, cricket, rugby, golf, running just a few of their many pursuits. We (Mum and Dad) covered countless miles ferrying them to school and county events. It would be wrong to see them as total angels: our son lacking navigation after a particular birthday party eventually rescued by his sister. Friday nights and all the gang getting glammed up for their trip to town – culminating in 'Bed Socks'.

A massive part of family life revolved around other family members and friends. Countless events over the years: memorable parties, christenings, birthdays, engagements, weddings, funerals and wakes – the list endless. Family and friends, as you may have deduced from the 'Sabbatical', a massively important aspect of our lives – and still so.

We've not got onto the topic of memorable family holidays – a number incorporated into prior 'Sabbatical' entries. Suffice to say, we've had an absolute ball over the years and now you grandchildren are involved too. As mentioned previously, the imagery of our COVID-19 lockdown walks demonstrates a place we value highly, and Granny and I look forward to enjoying this with you our family (children and grandchildren) for many years to come.

Authors & Books
Stuart Brown – Forbidden Paths

Stuart was born and grew up in Horncliffe. The strange aspect: I knew his Mum, the local publican, but only learned his story more recently. Taken prisoner prior to Dunkirk. Fell in love with a German girl. When hostilities ceased, brought her to the UK and married. Became editor of The Scotsman. Sourced by a favoured bookshop in Berwick, 'Slightly Foxed'.

Musicians & Popular Music
Heather Small – Proud

A big favourite from Granny's playlist. Great lyrics and profound: What have you done today to make 'You' feel proud? A question we could all benefit from if asked on a daily basis.

Composers & Classical Music
Ludovico Einaudi – Nuvole Bianche (Live – Steve Jobs Theatre)

Walks along the banks of the River Tweed wouldn't be the same without our faithful hound trotting along beside us (see final image). Einaudi's thought-provoking piece an apt composition for our final entry. Now firmly logged as Rory's (our new grandson's) classical piece.

Fine Wine
2006 Dom Perignon, Rosé Champagne, France

Having a son (your dad or uncle) whose occupation involves fine wine has benefits. On asking which wine we should conclude the Sabbatical with, he proffered a 2006 Dom Perignon Rose Champagne. Ostensibly to mark the birth and wet the head of our new grandson – so a perfect choice. How could we refuse, and absolutely delicious.

EPILOGUE

To Our Grandchildren

Granny and I set off on our 'Sabbatical' journey 53 weeks ago at a time of a national lockdown due to the pandemic. Our aim and intention was to generate a bit of light relief for family and friends as well as leaving you (grandchildren), with a brief insight - if we were to succumb. My opening letter explained my grandparents died before I was born.

Compiling the 'Sabbatical' has been an interesting exercise: at the outset a daunting task, but a lesson learned; break things into bite-size pieces (1,000 words a week rather than the 52,000 required) and your goal is attainable.

Unable to travel beyond our immediate locality made the memories of all the many trips we've embarked on over the years special - as well as proving a joyous and vicarious means of escapism. Plus an opportunity to engage with those who shared these special times – and to demonstrate how appreciative we are of family and friends.

Granny and I have completed our 'Sabbatical' trip of 52 destinations (both home and abroad), each having left an indelible mark. So we wrestled with how we should conclude and what our final communique should contain. How could we back up our original statement to you (our grandchildren), and one to which Granny and I very much subscribe:

The World is Your Oyster; Do with it What You Can – With an Open & Enquiring Mind

Broadening one's horizons via travel as well as meeting and engaging with people from as wide a range of race, creed, colour, background, and views - something Granny and I believe **are** of value. Playing safe and limiting one's dimension and view of the World is likely to prove constricting –widening one's perspective adds to life's rich tapestry.

Hence our decision to conclude with two further books, classical composition, piece of popular music and fine wine. All have played their part in the 'Sabbatical' and our aim to expose you to a wide range of topics and opinions. If any make you stop, think, contemplate, and examine others points of view and perspective – we've succeeded.

Authors & Books

Billy Connolly – Windswept and Interesting

Billy's story from Glasgow shipyard worker to musician, comedian, TV presenter, actor and artist on the World stage. Abandoned by his mother at an early age, plus physically, mentally and sexually abused during childhood, but look at all he's achieved in life and far from finished. An inspiration to others in handling adversity. An amazing ambassador for those affected by Parkinson's or Prostate Cancer. My all-time favourite comedian.

Natalie Cumming – The Fiddle

Charts the true story of an inspirational Jewish family who walked from St Petersburg to Odessa (1,000 miles) to escape the Bolshevik pogroms. Settled in the UK and eldest daughter (Rosa) became a violinist with the London Philharmonic and later the Berlin Philharmonic. Detained on Kristallnacht and transported to Auschwitz and later Belsen. Testified at the Nuremberg War Trials and died not long after. The violin inherited by her inspirational younger brother (Sonny). The story came to light via the 'Repair Shop', our favourite Lockdown programme.

Musicians & Popular Music

Roberta Flack – The 1st Time Ever I Saw Your Face

Takes me back many years – 2022 marks our 45th wedding anniversary. Tells a story relating to Granny and I. If you're lucky enough to find a partner, soulmate and friend who can match up to your Granny – you'll have done well.

Composers & Classical Music

Henryk Gorecki – Symphony of Sorrowful Songs 2nd Movement

Dedicated to the memory and courage of Rosa Anna Levinsky: the words taken from those etched on the wall of a Gestapo jail in Poland by an unknown 18 year girl during WW2. A small insight to the horrors of Hitler's Nazi regime.

Fine Wine

Alter Ego de Palmer, Margaux, France - 2016

Purchasing 'En Primeur' relates to laying down a young wine in bond to mature and develop. This final wine will not see the light of day for a number of years. As Granny and I continue on our journey, wouldn't it be marvellous to think we could enjoy a glass with our eldest grandson (who'll be legally of age) in a decade's time.

Our final thought relates to a BBC radio interview listened to late last year. A mum lost her talented and inspirational 18 year old daughter to cancer during the pandemic.
To shine a light on the amazing attitude demonstrated by her daughter she'd launched a charity to encourage and help others. This based on her daughter's words of wisdom:

**Look Up,
Look Out,
Be Kind.**

**All our Love,
Granny & Grandpa**

P.S. – During our 'Sabbatical' journey we utilised quotes from 3 speeches marking significant events.
These relate to receiving my Old Age Pension, my parent's Centenary celebration, and a retirement dinner in my honour. We thought these may provide a further small insight to what makes the pair of us tick.

EPILOGUE

65TH BIRTHDAY SPEECH - 2017

For once being the oldest in the room has an upside – which relates to the fact, as of today, I am in receipt of my 'Old Age Pension'. I realise this may well be galling to many of you who are likely to be 68 or even 70 by the time you find yourselves in the same position - hence my decision to invest some of mine in advance on you, my friends, with a good dinner.

A few months back, my beloved wife and I watched a TV programme of another milestone – a 50th anniversary rather than a 65th, and related to the Beatles making their iconic album 'Sgt Pepper's Lonely Hearts Club Band'.

One of the tracks has a link to tonight's proceedings and written by Paul McCartney to celebrate his father's birthday, and famously asks if what life will be like "when I'm 64". Another short but apt verse: "Grandchildren on my knee, JJ, Lozz, and ?"

50 Years ago, when McCartney wrote his song, I was 15, my career at a top Scottish Public School heading for a premature end, and a fairly wild and decadent decade beckoned. At the time I'd no concept whatsoever of being 65 or becoming a pensioner. My conduct over the next 10 years led Mum and Dad to believe the likelihood of me achieving such a milestone was slim.

It was the 60s and 'Making Love Not War' had more appeal than cooped up in an all-boys Public School. An appetite for wine, women and song - as well as an addiction to 'speed' (I should clarify this relates to cars and motorbikes, not drugs) may well have propagated my long-suffering parent's belief my standing here today was likely to prove a challenge.

I presume the question you want answered is: "How did the transformation from reprobate, to this model upstanding citizen, come about"? The answer relates to three things:

The first, a visit to one of my many local hostelries: 'The Highlander' near Ponteland, on Tyneside – where this very pretty, blonde Geordie Lass decided to introduce herself and chat me-up. The rest is history, and my No.1 wife of 39 years (our anniversary being on Saturday) transformed my life. No Internet dating back in those days.

The second, our family, of whom we are extremely proud: bringing children into the world is an honour, privilege and a responsibility. One has to learn the art of looking out for others rather than yourself. Luckily our children inherited their mother's genes and, as per her exemplary example, have proved to be paragons of virtue and a major influence on my rehabilitation. The addition of a No.1 son-in-law and No.1 daughter-in-law, as well as a No.1 grandson and No.1 granddaughter has added a whole new positive dimension to our lives.

The third relates to friends: we've friends we can have for dinner, other friends we can spend a night away with, and the most important friends are the ones who can put up with us for a week, 2 and more. You're very much the latter, and an important aspect of our lives.

There have been so many memorable trips and veritable feasts: Florida, Portugal, Lymphail, Glen Tanner, Cabo, Aya Tia Roa, Morocco, Croatia, Dordogne, Saddell, Monte Casino, Affric, Annie Girl, Glen Lyon, Ro-Ro, the Ness, Melbourne, Cradle Mountain, Queenstown, Dome Hills, Stuart Island, Samoen, Covara, Tobago, New York, the Lake District, the Laxford, and many more.

I'd like to take this opportunity to acknowledge your patience and help in assisting my wife in developing my skills as a 'Professional House Guest' – this is much appreciated, and has played a vital role in my ongoing quest to become the model citizen she so desires. I should add my apprenticeship still has a long way to go.

Great friends of ours use a phrase to which I totally concur: 'Families that play together, stay together' – to which I would like to add the word 'friends'.

If you would be upstanding I would like to propose a toast:

'To family and friends, and to lots more playing together in the years ahead'.

CENTENARY SPEECH – 2020

For many years when in residence, there was a ritual of visiting the family home on Friday evenings at 7:00, to have a drink and catch up with Mum and Dad. This originated back to us returning to the Borders – Christmas Eve 1982. Our family ritual continued for many years, all the way to the 24th November 2008 – when Dad sadly passed away.

We were very lucky to spend many happy evenings with the pair of them, and after Mum's death on the 13th October 1995, the ritual continued. Dad and I latterly talked about many things – past and present. On various occasions, he stated: 'You've done a better job in bringing up your children than us'. Something to which I disagreed - and we discussed the merits of this at length.

They were an amazing pair, and their story belies Dad's statement. Mum was born on the 4th September 1920 in Newcastle-upon-Tyne; Dad on the 23rd August 1920 in Berwick-on-Tweed. The year 1920 is significant, and the reason we're gathered here today - to celebrate the centenary of their birth.

Mum and Dad loved a party, and many memorable events and celebrations took place at our family home during their period of tenure, from 1950 to 2008. So it's highly appropriate we're gathered here today: to toast them, remember them, and give thanks for lives well lived.

Living through a time of a global pandemic, it's easy to feel sorry for oneself and dwell on personal problems and issues. Whenever a hint of such thought appears, I only have to relate back to Mum and Dad, and their generation, and all they endured with such stoicism, strength and resilience, to realise how lucky I and my generation have been, with any issues paling into insignificance.

Dad led me to believe he and his sister led a rather idyllic life up to June 1933, when tragically their mother, whom they both adored, died. My understanding is Dad was not informed or brought back from boarding school until after his Mum's funeral. Never one to feel sorry for himself – but imagine the trauma of such an event on a 12 year old boy.

At the outbreak of the 2nd World War in 1939, Dad at 19 was working in Sweden learning the timber trade, with a view to joining the family business. I never managed to get the full story; but am led to believe getting back to Britain proved quite an adventure as chaos ensued.

The next 5 years were spent involved in a war - of which only latterly did Dad speak. A defining moment in his life related to becoming a Gurkha officer – this totally changed his view on life and people. He possessed an amazing karma and approach to things, which came from his experience and time with his beloved Gurkhas. To the very end, his annual Gurkha reunion meant a great deal.

After Dad's death, my brother and I, plus wives, paid a nostalgic visit to Monte Cassino in Italy where he'd fought. The Allies suffered 55 thousand casualties during this infamous battle. The scale and enormity of what took place is mind-boggling for our generation. Dad suffered nightmares to his dying day as to the horrors they endured. But we, his family, never saw or witnessed these.

One last reference to Dad's war: for those who watched the 75th celebration of VJ Day (Victory over Japan) earlier this month. After Italy and Greece, his

regiment - the 7th Gurkha Rifles - were back in India preparing to be part of the main assault force and landing on Japan. He always said the atomic bombs, dropped on Hiroshima and Nagasaki, saved his life – finally bringing the War to an end.

It's worth bearing in mind the 3 generations here today may well not have existed but for this.

His return to Berwick after the War was marred by further tragedy, in that his father died on Christmas Eve 1946 - less than a year after Dad's departure from India. For all who knew Dad, you'd never have known anything had ever caused him pain or issues during his life – an amazing example to us all.

The love of Dad's life appeared on the horizon sometime after this, and became his beloved wife in 1948 - following a long courtship. To understand Mum, it's important to realise she also suffered tragedy early in her young adult life.

CENTENARY SPEECH – 2020

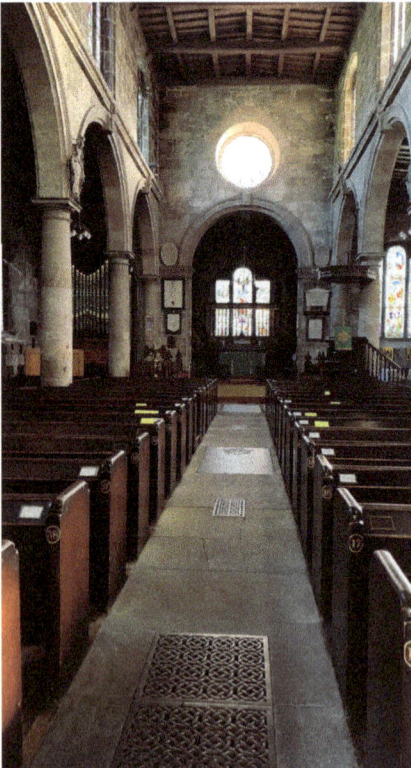

During the war Mum, a young WREN Officer, married a dashing young Naval Officer on the 11th April 1942. Jack was a submariner and, tragically, his vessel 'Thunderbolt' went missing after an operation off the Italian coast in March 1943 – less than a year after their wedding.

Mum's involvement in naval intelligence meant she was aware 'Thunderbolt' was missing, but as the enemy were not - she was unable to share this with others until later. At 22 years old, one can only imagine the heartache, sorrow and grief this must have entailed. Up to the ending of hostilities in 1945 she always hoped, beyond hope, that Jack was a prisoner of war – but not to be.

Dad recognised Mum's love for Jack - but would tell me, 'Over time she also grew to love me', and in his memoir he states: 'We grew together slowly'. They developed an amazing bond and understanding - possibly formed by the sorrow, grief and loss they'd both encountered in their younger days.

Sadly Mum, as per Dad, lost her father late on in the War to illness. I mention this because I and my siblings, other than my eldest sister, never met any of our grandparents - as Mum's much-loved mother Gladys died in 1949. Bear in mind our parents didn't have the support of grandparents in raising us.

I and my siblings were brought up in the family home in Horncliffe. Ours was a lifestyle that parents of today would no doubt frown upon. We ran wild, as was the norm in those days. It's also important to bring in the significance of the Black family. Our 2 families totally intertwined and spent as much time in one residence as the other.

We were extremely lucky. We may not have had grandparents, but Mum and Dad had an amazing network of friends who became our extended family: Joyce, Eric, Betty, Michael, Margaret, Jimmy, Doss, Jim, Midge, Geoff, Pam, Grant, Eva, Torquil - to name just a few. This was the norm to us, and we benefited.

For our time, the freedom we were provided was unusual – our parents had a more relaxed view and attitude than many of their generation, and I for one took full advantage.

It would be remiss not to mention Mum and Dad's amazing sense of duty, honour and responsibility. Their records speak for themselves: Dad ran the family business extremely successfully for 40+ years, he was a Magistrate, Tax Commissioner, Sheriff, Harbour Commissioner, Church Warden - to name just a few.

Mum's involvement with Polio, the Spastics Society, Save the Children, Meals on Wheels, the Church - and many others - aptly demonstrates her commitment and love for others. Their memorial services bore this out, with packed churches on both occasions and 100s of letters of condolence.

Sadly Mum suffered ill health over the years – this involved cancer and later a stroke. But she didn't believe in giving in to anything, or showing weakness. Her strength of character and stoicism were something to behold. Despite illness, it's important to point out Mum's latter years were some of the happiest of her life – she adored her grandchildren, as did Dad. You made the pair of them so happy and extremely proud. I'm sure they're watching over us here today -with the added bonus of having their 'great grandchildren' in attendance.

So: returning to Dad's statement, 'You've done a better job in bring up your children than us'. This a moot point – but if there is any truth whatsoever in this, they were the inspiration to making this possible.

RETIREMENT WORDS – 2021

I thought I may be called on to say a few words. Unlike Paul, the Master of public speaking and performance, I require time to think about what I want to say, and to put things down on paper. Hence my pre-prepared words.

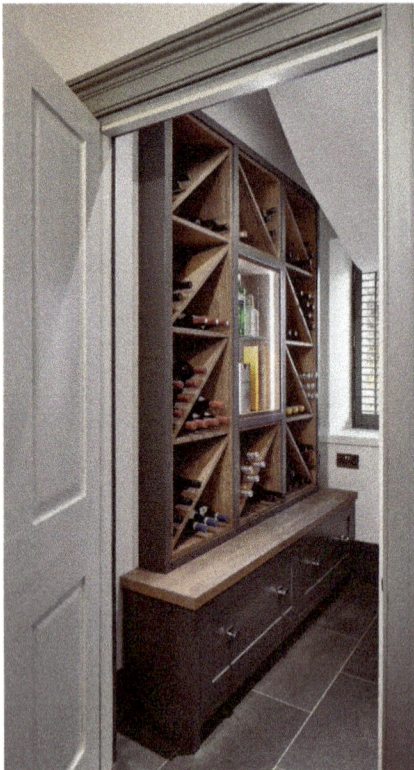

I was humbled and moved when receiving your invitation to a 'Retirement Celebration' – to honour me. I've never been one to court personal recognition and the limelight, the power of Team rather than the Individual something I've subscribed to for many years. KI the great example –hence recognition by you extra special.

My involvement with Paul and Gerry dates back to C. & J. Brown days. We worked with others to help the unfortunate individuals who'd lost deposits or had incomplete kitchens. This demonstrated undertaking what was right, and doing the honourable thing. So when the pair of them decided to set up in business at Stephenson Road, we'd no hesitation in agreeing to work alongside them. With no concept, KI would become the retailer others should emulate.

I could witter on for a long time about the past 27 years and our company's relationship and involvement with KI - but much of that's in the public domain. Some experiences such as the Sticky Wicket, lift excursions at the KSA AGM, memorable Awards Nights, the Buying Group and Crief Hydro events, Dealer Forums, KBB Exhibitions, and countless others - possibly better glossed over. Many, many memorable nights, and subsequent hangovers.

So, what could I possibly talk about? I thought I should try and impart what little wisdom I've gained from 50 years of work – 40+ involving the kitchen industry. I should begin by explaining Paul takes great delight in telling me I've learnt everything from him. I should stress, I gleaned the odd thing in the 40 years prior to his tutelage.

Part of my education took place less than a mile from here, at one of Scotland's prestigious Public Schools – almost 3 years from the age of 13. The less said about this the better, as my time at the institution came to a premature end.

I then went up market to Ashington Tech in Northumberland - the largest mining community in Europe at the time -

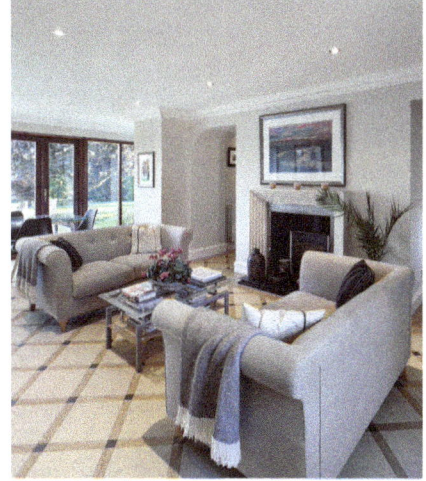

studying A-levels - which I failed, due to spending more time working in a garage in Newcastle than attending College. One beneficial aspect relates to doing a Technical Drawing course – this proved most useful later in my career.

One of the few positive aspects of my time in education took place just up the road from here, at Basil Patterson's. An educational establishment for rejects. I spent 2 terms cramming for A-level resits. A remarkable lady taught me – Mrs McDowell. My first tutorial involved précising a piece and being duly informed I'd no possible chance of attaining an A-level – but she "would teach me to appreciate English, marshal my thoughts and to put these down on paper".

One of the most important lessons of my life – although a decade later, before I got the tools to put this into practice. The Amstrad Word Processor came to market, opening up a whole new world for those who couldn't spell. I was most likely dyslectic,

not something recognised all those years ago. I should add I shocked everyone by passing my exam, enabling me to enrol at Newcastle Poly to study business – which proved Dickensian, so I left after 2 terms.

At 19, in the early 70s, I went to London to become a Management Trainee on Tottenham Court Road, in Europe's first Hi-Fi Department Store. An excellent grounding led by a Scot, who'd previously run the Pettigrew & Stephens Department Store in Glasgow, alongside some high-flying M&S management. I had a ball, as well as loving my work.

I became Store Manager for their Nottingham branch at 21. Subsequently poached to set up a new operation for a Northern department store. Naively went solo and set up my own business, and lost everything at the age of 24. Hence became unemployable, and no option other than to plough my own furrow.

You may well ask, what came about to turn me into the model upright citizen you see here before you? While having a quiet pint in a hostelry on the outskirts of Newcastle, I was picked up by a very fetching, young blonde Geordie nurse, later to become my wife. She's worked extremely hard over the last 44 years - attempting to bring me to heel.

From a Property Development and Construction Company, set up in partnership with a friend – later morphing into M/H Manufacturing. I've hopefully provided an insight to my career prior to kitchens.

RETIREMENT WORDS – 2021

So what have I learned over the last 50 years of my working life?

Negativity the preserve of the masses – Originality & Positivity a rarer commodity.

Everyone can tell you what you can't do – be willing to stand by your Judgement & Dream.

Health & Happiness inevitably outdo Wealth & Status - in my experience.

Develop a Work-Life Balance incorporating Family & Friends - something I subscribe to.

Learn to Listen, leave a meeting with a view opposite to that which you entered - uplifting.

Simple Good Manners and an ability to say Sorry when duly required – a powerful tool.

A Well Done or Thank You in appreciation of a job well done – of immeasurable value.

There's no I or Me in Team. The power of a Positive & Capable Team - immense.

Without a focused Aim & Goal + a subsequent Strategy & Action Plan – likelihood of failure increases.

Drive & Passion, and an ability to impart these - critical ingredients for a successful business.

Engagement with Education & Learning to eradicate issues in advance – pays major dividends.

Communication Skills and the ability to develop Meaningful Relationships – vital in the World of Work.

Building Trust & Partnerships – important ingredients throughout my career.

Input & Effort without achieving a Result – tends to be a waste of resource.

Identify your Strengths & Weaknesses – and work with others to mitigate your frailties.

From a management perspective – Respect & Creditability more important than being liked.

Humour & Humility – play their role in resolving issues and engendering a sense of fun.

Ensure your Word is Meaningful, if you say you're going to do something - make sure it happens.

An imbedded positive can-do Ethos & Culture throughout an organisation – works wonders.

The Development of Friendships with Industry Colleagues – has played an important role in my career.

I realise I'm preaching to the converted.

The vast majority of my 'words of wisdom' have been enacted within KI over many years. Hence I salute you all: it has been a massive privilege, honour, pleasure, and education to have worked alongside the KI Team for so many years. Once more, thank you for this very special occasion – it means a great deal to me.

I'd like to finish by proposing a toast: to KI, and your ongoing success.

70TH BIRTHDAY BASH - 2022

To many, the act of celebrating their 70th birthday could well be viewed as an oxymoron. But taken in the context of doing so in the company of a select group of 'Sabbatical Friends' alters one's perspective.

To comprehend what I'm attempting to impart requires an explanation of 'Sabbatical Friends':

They're like-minded people who have the good grace not to take themselves - or fellow members of our illustrious group - too seriously. They have the ability to accept and laugh (politely) at their own and other 'Sabbatical Friends' eccentricities and foibles. They're comfortable in their own skin and know how to relax in the company of others. They can cope with the light-hearted banter which inevitably takes place whenever we're in each other's company.

Basically they're great mates who know how to enjoy themselves and have a good time.

Sadly, on this momentous occasion, it's not been possible to gather all our 'Sabbatical Friends' together - as various members of our illustrious band live on the other side of the World. But those unable to attend are in our thoughts, and look forward to a belated celebration when next we meet.

As many of you know, I compiled a journal (The Sabbatical) during the 2nd COVID-19 lockdown, directed at our grandchildren. This highlights a great many of the trips, adventures and magical mystery tours embarked on over umpteen years. A further aim was to point out to the youngsters some of the many experiences, possibilities and opportunities open to them in life.

The 'Sabbatical' is to be published early next year, so I'm not going to regale you with the many tales incorporated. But we'd like to take this opportunity to thank each and every one of our 'Sabbatical Friends' for great times and memories - without you, none of these would have been possible.

Back to the task in-hand; celebrating my 70th birthday.

It would be easy to look at the occasion in a negative fashion and lust after being young again. But without reaching my

8th decade and all the many experiences encountered along the way, I would be a lesser person and lack the numerous benefits and knowledge gained over the intervening years.

45 years with my long suffering No.1 Wife emphasises the passage of time. Without the Matriarch, our much-loved daughter and son wouldn't exist. They've since married amazing partners, and between them brought into the World our 4 adored grandchildren. For this to have transpired requires copious years to pass, and hence a reason to celebrate - how much poorer life would be without them.

The pair of us often pinch ourselves and discuss how lucky we've been in life. From academic rejects who started married life without a bean – we've done OK. We live in a very special part of the World (the English/Scottish Borders), the home we've built and nurtured is very important to us, we've both created businesses and revelled in the challenge, and lucky to have enjoyed good health.

Beyond this, and by far the most important achievement, relates to the importance we place on relationships with family and friends – for us, what's the use of longevity without these? So I'd like you to all be upstanding and drink a toast:

To Family & Friends & More Good Times Ahead.

One Last Sabbatical Book & Author

The Shoes of A Foundling – Mink van Rijdijk
English translation by Anneke van Deurse-Verhave (Rose's friend, and Mink's cousin).

Recently I was given a book by a very dear friend (Carol) who has embarked on a number of 'Sabbatical' adventures covering a great many years. It's a particularly poignant read in that Carol and Rose (whose story this is) have formed a bond, in no small part due to traumas they've both encountered. 'The Shoes of a Foundling' relates to Amsterdam's own Kristallnacht and the despicable treatment of Dutch Jews by Nazi Germany and the Dutch citizens who conspired with them.

Rose (Roosje Drukker) was given up by her Mother (Fietje) as a foundling to save her daughter's life. The book details Rose's life long battle with the ramifications of this: the utter despair and mental torment over many decades brought into focus much later in life by a series of what she refers to as 'Miracles' – events which have brought about a semblance of peace and understanding.

Fine Wine
Laurent-Perrier, Grand Siecle, Iteration NO 24 MV

Consumed in the company of our No.1 Son and No.1 Daughter (supplied by them) prior to a memorable dinner to celebrate our 45th wedding anniversary.

PHOTOGRAPHY

The vast majority of the photography incorporated into the 'Sabbatical' is via my own fair hand. But due to the pandemic, travel restrictions, weather and other unforeseen circumstances, I was unable (in a number of instances) to get vitally important images. Hence I would like to acknowledge and recognise the individuals and organisations that helped me rectify such situations.

Week 18 - Ford & Etal (Northumberland). Due to Lockdown and Government decree it proved impossible to get an image of Bunty (the train) as the miniature railway was unable to operate. I would like to thank James (Lord Joicey) and the Ford & Etal Estates Team for providing this image.

Week 28 – Lochinver & Assynt (Scotland). I mention in the text that Suliven was enveloped in cloud at the time of our visit/ trek – hence an image provided by my long-time friend Carol.

Week 31 – French Riviera, Majorca & Amalfi Coast. I failed with a suitable image of Stalca, hence this supplied by Janet (a great mate) whose magnificent vessel this was, as at that time.

Week 43 – Provence (France). When compiling this journal entry I mention Entrecasteaux and the Gorge du Verdon from decades ago - and no photography available. Luckily my sister-in-law (Patsy) and Cousin (Michelle) came to my rescue.

Week 44 – Tintagel, Launcells & Moretonhampstead (England). No imagery of my Nephew's wedding, luckily Toby Butler (the official wedding photographer) came to my assistance – thank you.

Week 46 - Lake District (England). For some inexplicable reason I could find no suitable photography of the High Fells, so once again my sister-in-law (Patsy) provided a couple of cracking images.

Week 47 – Dalaman Coast (Turkey). At the time I wrote this particular entry we'd not as yet sailed on Salamander - our trip yet to take place. I would like to thank Tessa and Salamander Voyages for the 2 images of their magnificent vessel – also to report back on a subsequent spectacular holiday.

Week 48 – Edinburgh (Scotland). I was mortified on visiting the Botanical Gardens to find the Palm Houses out of bounds due to restoration work. Luckily a long-time friend and great photographer (Michael Barron) had images of these and willing to let me incorporate such into the 'Sabbatical' - also his great night-time shot of Murrayfield (much better than mine).

Week 50 – London (England). At the time of this journal entry COVID infection rates sky-high and the weather not advantageous for photography. Hence I enlisted the assistance of a couple of mates. Once again Michael Barron helped out with an image of Kew Gardens, and Darren Chung (a good friend and professional photographer) came to my rescue with others. I must also thank English Heritage and their photographer Anna Kunst for the image of Chiswick House.

Retirement Dinner Words. I thought it apt as this relates to work, demonstrates a few kitchens we've manufactured. These images taken by Darren Chung, my go-to photographer for 20 years.

For details of new and forthcoming books from Extremis Publishing, including our monthly podcasts, please visit our official website at:

www.extremispublishing.com

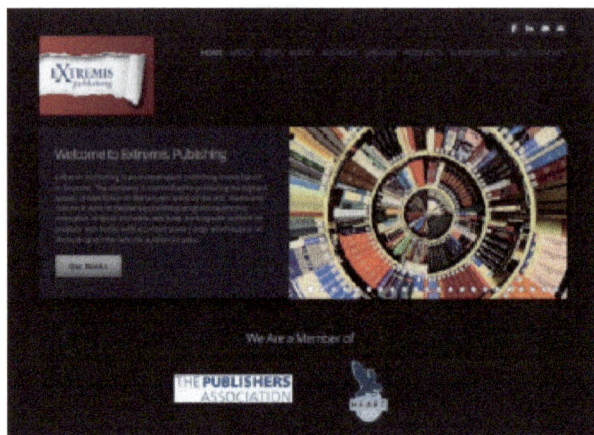

or follow us on social media at:

www.facebook.com/extremispublishing

www.linkedin.com/company/extremis-publishing-ltd-/

Lightning Source UK Ltd.
Milton Keynes UK
UKHW051155291222
414539UK00004B/19

9 781739 854362